PACKAGING FOR THE UNENLIGHTENED

By

James B. Vandegrift
J.V. Technologies

ISBN: 1-4107-3394-7 (e-book)
ISBN: 1-4107-3393-9 (Paperback)

This book is printed on acid free paper.

1stBooks - rev. 05/15/03

ACKNOWLEDGEMENTS

I wish to extend my deepest appreciation to Romarc Corporation and especially Joan Saalfrank and Dean Marino. Without their urging and confidence building, this book would have never been completed. I also wish to thank them for making it readable and in compliance with proper grammar and punctuation.

I also wish to express my sincere gratitude to Joe Thomas for his depth of knowledge in hazardous material packaging. As the manager of packaging engineering at a major automotive battery company, Joe had to master the intricacies of hazmat packaging on a daily basis.

To the many people in the packaging field who over my forty five years of experience taught me much of what I know.

INTRODUCTION

Familiarity with packaging materials, manufacturing styles, dimensional sequences, costs, designs, and pricing is not difficult to learn. The problem is, some packaging suppliers would like to keep you in the dark and dependent upon them.

My intent, and the purpose of this book is to provide you with down to earth useable knowledge. To help you weed out bad suppliers and give you an opportunity to lower costs. To provide you options in package design. To make you aware of different cushioning material, equipment and procedures. To put you on a level playing field with present and future suppliers. To convey to said supplier that you are informed and knowledgeable and not a pushover for his/her agenda. To make you aware of the intricacies of government and hazardous material packaging.

Packaging purchases are, in many companies, low on the priority lists. Many dollars are expended that could possibly increase the bottom line. Your suppliers are controlling the amount of money you spend. This book should help you spend it wisely.

The majority of people involved in packaging are not graduate packaging Engineers. Usually they were assigned the task of solving a current packaging problem or given the responsibility for packaging the company's products. In either case, they were unenlightened and at the mercy of current and potential suppliers. The school of hard knocks can be rough.

CONTENTS

CHAPTER 1

CORRUGATED CONTAINERS

IN THIS CHAPTER

PURPOSE OF CORRUGATED BOX

The corrugated box provides the following, namely:

A surface to imprint data on such as handling instructions, ship to, shipped from, advertising and warnings.

A means of consolidating a quantity of pieces in a single enclosure.

A means of enclosing an item along with protective inserts or materials.

A means to provide stackability for small or irregular shaped products.

The corrugated box itself provides very little in protection against shock. The inserts/cushioning materials within the corrugated container provide the bulk of the shock dampening. Interior packaging comes in many forms and materials. Corrugated companies design inserts in corrugated only. Foam companies (molders/fabricators) design inserts in their product. Likewise many companies that sell cell type partitions, bubble wrap, honeycomb, foam-in-place, or bulk (peanut) materials tend to push their product over other solutions. It is rare to find someone or some company that is familiar with all of the above to the extent that a composite package can be designed with optimum protection and at a reasonable cost. Most sales representatives have a working knowledge of their product only. Sometimes a shipping supply house can utilize its sources of supply to arrive at a composite package. You or someone selected by your company could also coordinate a composite package. Many people have got their start in packaging by being given an assignment to look into a packaging problem. This is called learning in the school of hard knocks.

SOURCES OF CORRUGATED BOXES AND MATERIALS

Boxes and/or materials can be purchased from five basic sources:

1. Corrugater
2. Sheet Plant
3. Broker
4. Stock House
5. Shipping Supply Company

CORRUGATER

This source does everything. It makes the corrugated board. It converts the board into boxes or inserts. It makes pads and sheets. This source usually has an extensive customer service section, which can design and provide samples. Some are national companies, which own forests and make their own paper.

SHEET PLANT

The sheet plant purchases the corrugated board from the corrugater. The sheet plant converts the board into boxes or inserts. They will purchase sheets and pads from the corrugater and re-sell to you. The level of customer service can be similar to the corrugater or almost non-existent dependent on the company.

BROKER

The broker makes nothing. The broker farms out your specifications to various sources for pricing. He selects the lowest price, adds his mark up, and quotes you.

STOCK HOUSE

As the words imply, the stock house stocks many styles, sizes, and test values, which can be purchased in small to large quantities. Usually, the price is F.O.B source and you pay the freight charges.

SHIPPING SUPPLY COMPANY

This source stocks various shipping supplies such as tape, bubble wrap, plastic wrap, carton seal tape, some boxes, etc. They make nothing. They will buy and sell special boxes like the broker.

Note: Corrugaters, sheet plants and brokers usually price their products with delivery charges included. Stock houses do not include delivery in pricing. They ship “freight collect”. Shipping supply companies usually include delivery in pricing. Some specialty materials such as fabricated and molded expanded polystyrafoam are shipped freight collect. Obviously, when comparing prices from different sources, delivery costs must be included.

MANUFACTURING YOUR BOX

Corrugated boxes and inserts are made using Kraft paper. There are variations of paper based on its weight in pounds per 1000 square feet. These weights are defined in “materials” section. As the box puncture tests vary, the weight of paper used also changes. Rolls of paper, to the desired weight per 1000 square feet, are combined on a corrugater. Here the liner and medium papers are glued together to form corrugated board. The width of the corrugater and the resultant overall width of the board it produces can vary depending upon the source. Corrugaters can vary from 69 inches to 103 inches or more. The useable portion of the combined board is approximately 3 inches less due to misalignment of paper rolls. A 69-inch corrugater will have approximately 66 useable inches. The width of the corrugater (69" to 103") is also the direction of the flutes. Inasmuch as the flute direction is also the height (depth) direction of the corrugated box, the combination of depth plus top and bottom flaps equals the amount of material required within the usable width of the corrugater.

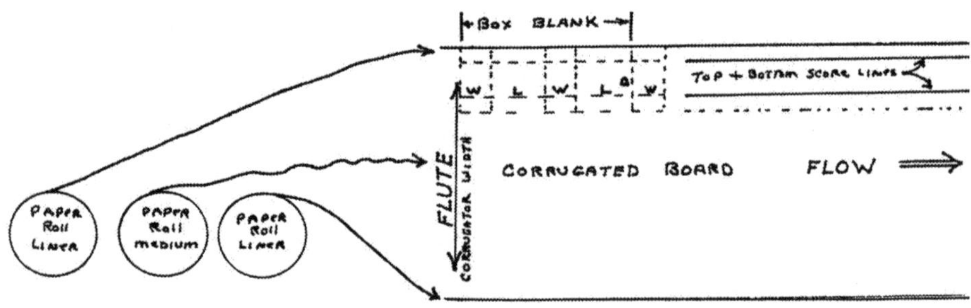

The owner of the corrugater will strive to maximize the full width of the useable board. He will check different combinations of orders before scheduling to maximize profit and minimize waste. If your order is delayed and you are advised that the delay is due to trimming out, they are losing too much material off the corrugaters to waste. Hopefully another order will combine with yours to more fully utilize the useable board width.

To provide a means of bending the corrugated board at a precise location, rollers imbed a "score" line into the material. These score lines permit bending the top and bottom flaps of the box and also the four corners of the box.

The corrugater is capable of providing the box's top and bottom score lines. These score lines are perpendicular to the flutes in the board and parallel to the flow of the board. Note: The four box corner score lines are added in the next operation. The corrugater also slices and cuts the board to the blank dimensions required for your order. Trim sheets and pads are sliced and cut to size without score lines and are bundled for shipment to you. Small pads may require additional cutting due to size.

The blanks are moved to the next operation. If blanks are not die cut, they go to a printer slotter. It is here that the scores and slots for the four corners are added plus the manufacturers joint tab for glued or stitched boxes. If the box requires printing, it is done during this operation. There are three types of joints manufactured: taped, glued and stitched. Taped and stitched generally require a separate operation. If your supplier has a folder gluer attached to the printer slotter, and the box size will fit, the glued joint can be completed within the printer slotter operation. This causes the box cost to be lower. If your box size is too large or too small for the folder gluer, then your box will require a separate operation.

Box making equipment is usually identified by what it does. A "printer slotter" cuts the slots in the blank, which permits flaps to be independently folded over. This machine can also print the blank if required.

A "folder gluer" will run the blank thru a series of arms which in turn will fold a four panel blank into a two panel knocked down box. During this operation, glue is applied to the manufacturers joint tab. Knocked down box with glued joint is then stacked and bundled.

As in most things, there are dimensional limits on various types of equipment. Printer slotters come in various sizes. There are limits to blank sizes. Some suppliers cannot make your box when depth plus two flaps and allowances exceed 63 inches. When two box lengths and two widths exceed the supplier machine limits, the box has to be made in two pieces with two manufactured joints.

On boxes with multiple score lines, there is a physical limit on how close they can be without die cutting. The scoring "wheels" have a protrusion or boss that limits how close they can get. Some holes on sides of box can be obtained with a "cir-cut" tool added to scoring wheels on printer slotter. Example: Hand Hole.

DEFINING AND MEASURING SCORE LINES AND BOXES

Score lines on boxes are applied on the corrugater and the printer slotter. The corrugater applies the top and bottom score lines. The printer slotter applies the score lines at the four corners of the box. The scores are applied with two rollers mounted on the equipment. These two rollers form a score line with this cross section appearance:

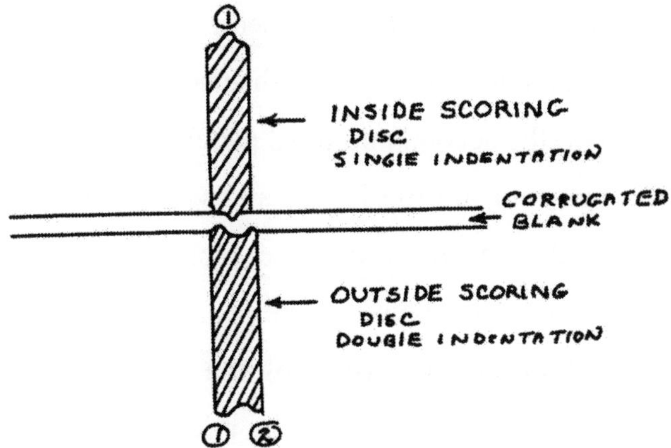

This creates on the box a single line indentation on the inner surface, and a double indentation on the outer surface of the box. Corrugated is always formed by bending toward the single line surface and away from the double line surface. This is very important when you are forming corrugated inserts. To measure distance between score lines, you measure from the center of the single line scores. Flutes can distort the actual bending line. You may have to look closely for the actual score line indentation.

Boxes are measured two ways:

1. Form the box and measure inside distance between sides of box. Depth dimensions are determined by measuring the overall outside height of box and subtracting four flute thicknesses.
2. Cut away manufacturers joint and open box to expose all four panels in a flat state with inside of blank facing

up (single line scores exposed). Disregard first and fourth panels. Measure the distance between the score lines on the second and third panels. Note: Panels one through four are the length and width panels. Panels two and three represent one each length and width. The dimensions between the score lines represent the desired inside dimension of the box plus a scoring allowance. On “C” flute, the allowance is 3/16 inch. On “B” flute the allowances is 1/8 inch. What this means is that to make a box 10 x 8 inches, you have to score the second and third panels in “C” flute 10 3/16 and 8 3/16 inches and in “B” flute 10 1/8 x 8 1/8 inches. To check the depth dimension, measure the distance between the tops and bottom score lines and subtract 5/16 inch for “C” flute and ¼ inch for “B” flute. This is the only true way to determine the supplier’s dimensions of the box.

MATERIALS USED TO MAKE YOUR BOX

Corrugated boxes are made from fourdrineer-processed Kraft paper. This differs from other papers inasmuch as the fourdrineer process has the fibers going in all directions, which provides better tear strength. Corrugated box papers are specified according to their weight per 1000 square feet. The most popular weights available are as follows:

26 pound	linerboard
33 pound	"
35 pound	"
38 pound	"
42 pound	"
69 pound	"
90 pound	"
26 pound	corrugated medium
33 pound	"

Note: corrugated medium consists of short fibers with strength in one direction.

Combinations of above linerboard weights are used to achieve the test values or ratings of corrugated boxes. These test values range from 125 pound test to over 1300 pound test. These values represent the hydraulic pressure or force needed to split (rupture) the boxboard under test. 125-test means paperboard, in test device, will split open at 125 pounds per square inch pressure. Examples are:

SINGLE WALL

Single wall corrugated consists of two linerboard liners and one corrugated medium. The weight per 1000 square feet of the facings controls the test values of the corrugated board as follows:

Linerboards	Test value
26# + 26#	125 pound
33# + 33#	150 pound
38# + 38#	175 pound
33# + 42#	175 pound
42# + 42#	200 pound
69# + 69#	275 pound
90# + 90#	350 pound
35# + 35#	200 test per Edge Crush Test (ECT) not Mullen Test

These test values are obtained on a "Mullen" test machine. Hydraulic pressure is applied to a one-inch diaphragm until the linerboard paper ruptures. The corrugated medium does not contribute to the test value.

Double wall corrugated has three liners and two mediums. Different paperweights are combined to achieve test values. Mullen type test not used. Examples are:

DOUBLE WALL

Linerboard weights	Test value
42# + 33# + 42#	275 pound
42# + 42# + 42#	350 pound
90# + 42# + 90#	500 pound
90# + 90# + 90#	600 pound

TRIPLE WALL

Triple wall corrugated has four liners and three mediums. Test values of 700 pounds to 1300 pounds are achieved through variance in linerboard weights. Flute sizes can vary but “A” flute is required on government specifications and is a popular choice in commercial triple wall.

Note: Test values DO NOT reflect permissible load weights for that particular box. Box maker certificate on carton flap will state permissible weight of load and upper united inch limits on box size. United inch limit and method of measurement varies. UPS united inches is outside girth plus box outside length. Basic box rules call for inside length plus inside width plus inside depth for united inches. United inches is a term developed to control overall box internal volume per particular test values. As the box volume increases, the test value of the corrugated board is increased because more box strength is needed. A large box of feathers could require a 175-pound test box because of weight but its "united inches" could require a 275-pound test box.

If no name and location appears on certificate, compliance of box is highly improbable! If you exceed the weight or size limits in the certificate, carriers will reject any damage claims! Although the certificate certifies compliance with Rule 41 of Uniform Freight classification for rail shipment and Item 222 of National Motor Freight classification for truck shipments, it is NOT A GUARANTEE of compliance. There are suppliers who will provide lower test material and certify a higher test value. Some will be known by reputation, others can only be found by a simple test. See testing for board compliance.

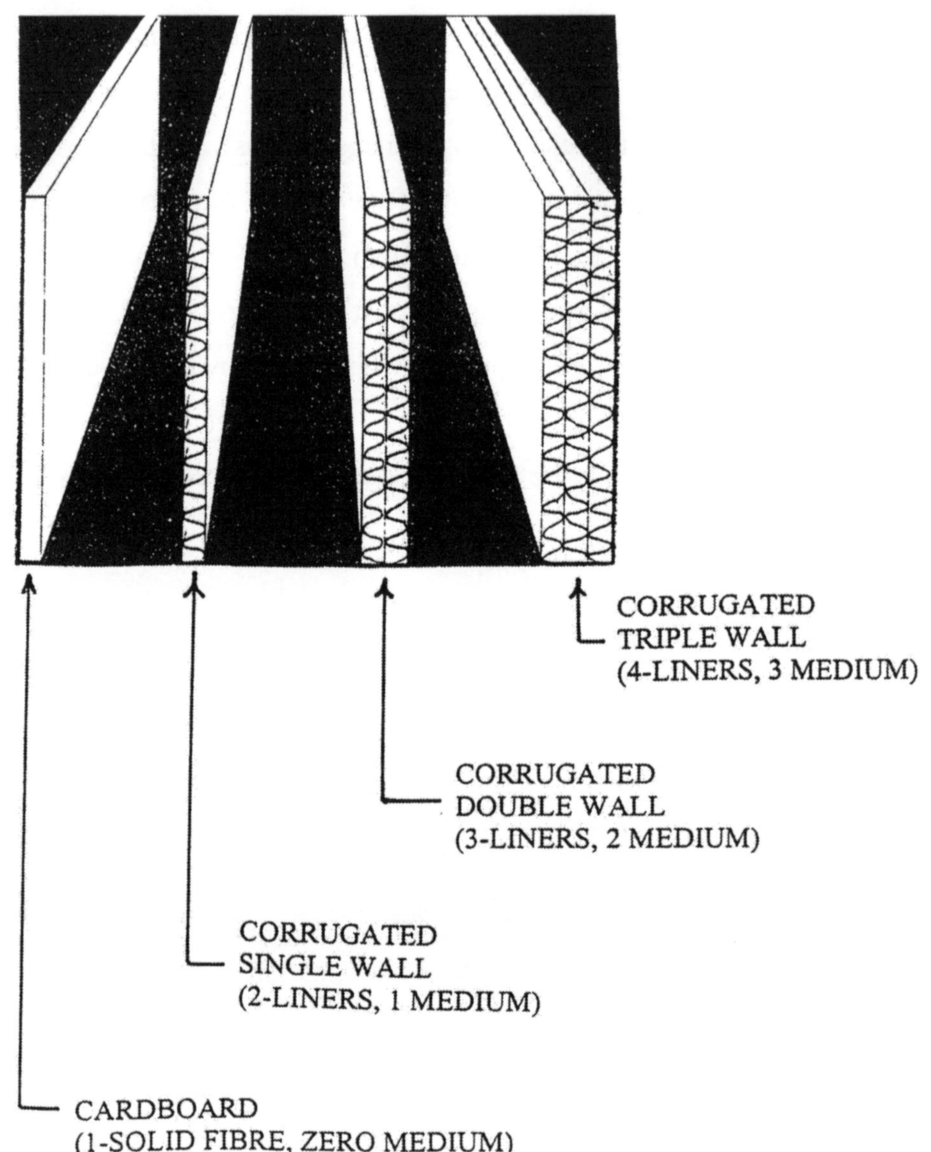
CORRUGATED
TRIPLE WALL
(4-LINERS, 3 MEDIUM)
CORRUGATED
DOUBLE WALL
(3-LINERS, 2 MEDIUM)
CORRUGATED
SINGLE WALL
(2-LINERS, 1 MEDIUM)
CARDBOARD
(1-SOLID FIBRE, ZERO MEDIUM)

CORRUGATED BOARD FLUTES

The corrugated flutes (the wavy paper between the outer and inner liners) are available in different sizes.

(A) 3/16 inch high
(B) 1/8 inch high
(C) 5/32 inch high
(D) ½ inch high
(E) 1/16 inch high
(F) 1/32 inch high

As the flute size changes, the undulations per linear inch change. 3/16" high flutes will have the fewest undulations and 1/16" high flutes will have the most undulations. As the undulations increase, the resistance to "flat crush" increases in die cutting. There are flute gauges available to check your box material. The steel corrugating rolls in corrugating machines that form the flutes can and do wear. This can distort the shape of the flute and make it more difficult to determine size. Never use edge of board to check flute size. Cut 2 to 3 inches away from edge to check flute size. Edges are usually crushed in process.

Flute size determines board thickness, which can affect cushioning, outside to inside dimension, dimensional layout to make a specific box or insert, plus resistance to flat crush.

DIE CUT BOXES

Die cut boxes also require a dimensional sheet off the corrugater, as previously described in this chapter. The sheets are taken to a die cutting machine. The sheet size can reflect one or more pieces to be die cut per hit. Example: a four out die will produce four pieces per each hit of the die. If the die cut box requires a manufacturers joint (tape-glue-stitch), the manufacturers joint will be done on a different machine. Printing on the carton will be done prior to die cutting, usually on the same machine with printing dies equivalent to the number of pieces in the cutting die. The die cut blank contains an extra ½ inch on all four sides as an add on to actual piece or pieces over all dimension. This allows some variation in blank placement prior to entering die cutting area of machine.

Die cut boxes can contain full tuck lids and automatic locking bottoms. They can have manually folded locking bottoms. They can contain holes, rounded corners, perforated cut lines, angled cuts or score lines, or other peculiar features. They can combine "RSC" or "FOL" tops with locking bottoms. See styles of boxes for meaning of RSC and FOL.

Insert corrugated materials with holes, curved edges, offset score or cut lines, multiple close scores or internal slots will require die cutting.

When boxes or inserts are die cut, waste material is usually produced within the piece. This waste material should be removed before the next operation or preparation for delivery. Extra steel rule may be added to cutting die to cut excess into small pieces to be removed by vacuum equipment or the excess must be hand removed. "Bobst" dies with male and female stripping dies are available on "Bobst" die cutting machines.

There are two basic cutting dies, namely flat and rotary. Rotary dies are made to fit different diameter drums.

On die cutting machines, the steel rule cutting edges on rotary dies leaves a serrated edge on the cut line of the box or insert. Rotary cutting dies are not transferable to another supplier unless the new supplier's machine will fit the die. Flat dies are more universal with overall size being the limiting factor. Serrated steel rule is not used on flat cutting dies.

Flat crush (the squeezing or flattening of the flutes in the corrugated board) is reduced in die cutting through the use of smaller flutes. Cutting dies contain resilient sponge-like pads to push the die cut piece off the steel cutting edges. These pads exert a lot of pressure on the surface of the piece being die cut. Flutes (A) and (C) do not resist the pressure as well as (B) and (E).

Die charges: see printing and cutting dies costs.

STYLES OF BOXES

(NO DIE CUTTING)

There are many box styles available. The most popular are as follows, namely:

a. Regular slotted box (RSC): length flaps meet in middle of box. End flaps do not meet.
b. Full overlap slotted box (FOL): length flaps are almost full width of box.
c. Five panel folder (FPF): corner cut, full flap five panel folder
d. Half slotted box (HSC): bottom of box open – no flaps
e. One panel folder (OPF): one panel folder

These boxes can be made in hundreds of sizes. See illustrations for graphic view of each style. The dimensional sequence of each style is etched in granite: it is LENGTH x WIDTH x DEPTH or L x W x D. The box opening (where you load the product), is defined by the L x W dimensions. Ordering the box with dimensions out of proper sequence will result in delivery of a box you can't use. Example: A box 20 x 10 x 8 will be much different than a box 10 x 8 x 20 or 20 x 8 x 10. The cubical content will be the same but loading the parts and inserts may be impossible.

Note: The width dimension can never exceed the length dimension.

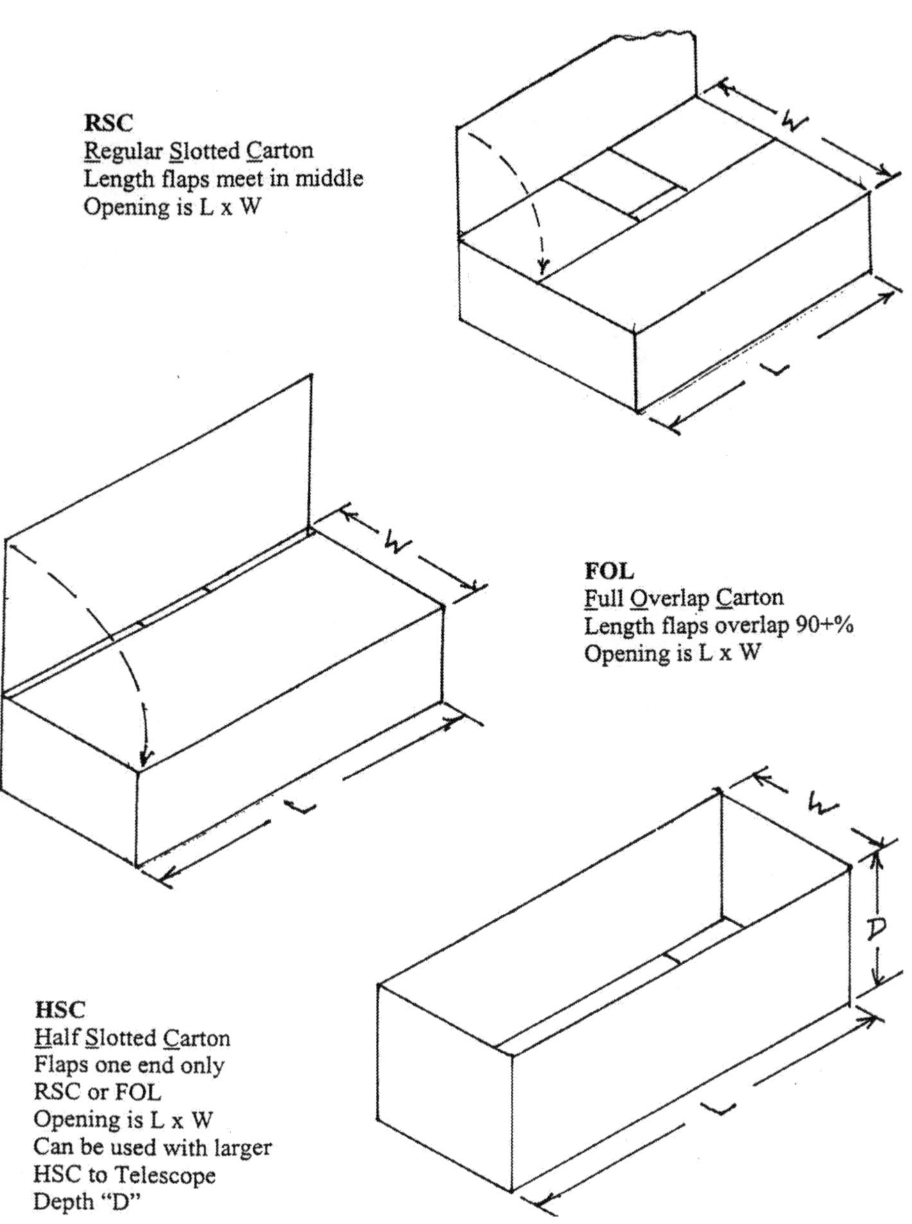
RSC
Regular Slotted Carton
Length flaps meet in middle
Opening is L x W
W
L
FOL
Full Overlap Carton
Length flaps overlap 90+%
Opening is L x W
W
L
HSC
Half Slotted Carton
Flaps one end only
RSC or FOL
Opening is L x W
Can be used with larger
HSC to Telescope
Depth "D"
W
D
L

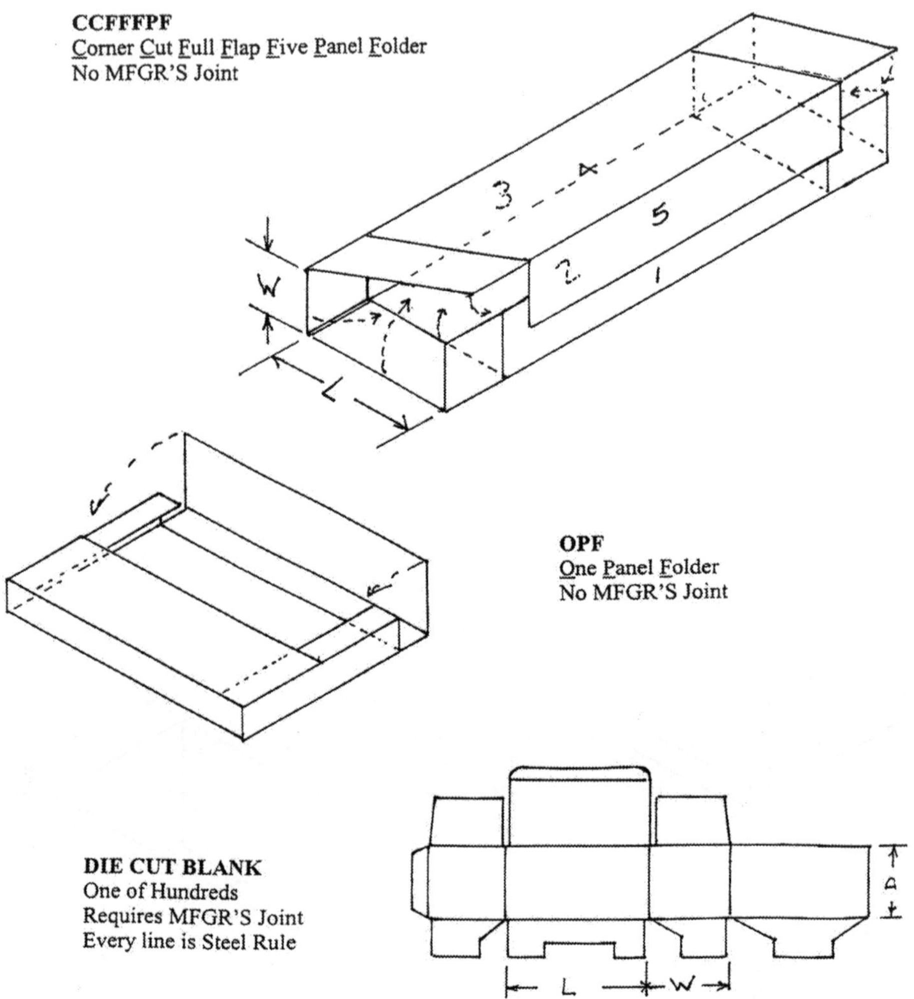

CCFFFPF
Corner Cut Full Flap Five Panel Folder
No MFGR'S Joint

OPF
One Panel Folder
No MFGR'S Joint

DIE CUT BLANK
One of Hundreds
Requires MFGR'S Joint
Every line is Steel Rule

Note: On RSC, FOL, and HSC styles, the width dimension controls material cost. The least material cost (per selected test) will be when the width is the smallest dimension.

PRINTING AND CUTTING DIE COSTS

Printing dies are usually made of rubber or poly. The poly dies reportedly provide greater sharpness on small letters and graphics. Printing die costs run close to $1.30 per overall square inch of die. The overall size of the die contains a 3/8-inch border on all four sides beyond the limits of the imprint. Fancy logos or special printing will increase the cost to $1.50 per square inch. The printing die can be a small portion of the imprint or the total surface of the box blank. Many print dies are for "up" arrows, handling data, ship to data, company name, address, etc. These small dies can leave significant surfaces blank. A large die would increase your cost by $1.30 per square inch of unprinted area. The choice of dies and die coverage is yours or the powers to be. Many corrugated companies mark up the costs of printing dies as much as 100 percent. Some do not mark up at all. When you pay for the die or dies, they become your property. You can leave them at your current supplier or transfer them to another supplier. The die costs are a one-time charge. Repair and replacement is the responsibility of the supplier from whom you purchased the dies. Transfer of dies voids responsibility by the original supplier. When warranted, some companies have purchased their own dies and loaned them to a supplier to use on their boxes. This eliminates the potential for the costly mark-up suppliers add on. If your company has facilities scattered across our country, a central source of printing dies can provide dies to all locations at a price you have negotiated and only one master plate was required per die. Your printing die source can run multiple pieces of each die which can lower costs and provide rapid replacement service to your plants. Most corrugated box suppliers want to pre-mount your dies to a backing sheet per box size purchased. This cuts their printing die set-up time. I have never seen a box quote containing a price based on mounted or unmounted dies. The cost benefit is 100 percent to your box supplier and zero to you. Oh yes, your supplier does charge you for the labor and material to mount your dies. If you have different size boxes requiring the same printing dies, extra dies will be required to provide separate mounted dies for each box size.

MOUNTED DIES – MULTI-COLOR

If you require multiple colors on your corrugated box (white not included), a separate set of mounted dies will be required for each color. Pre-mounting is necessary to provide proper registry of colors. An improperly registered set of dies can result in a very bad appearance.

COMPLEX COLORS, ETC.

On boxes where numerous colors and complex images are required, a "Litho" covered box may be required. In this situation a pre-printed Litho is separately run on special paper by a printing company. This paper, when required, is then mounted (glued) to a box blank. The combined box blank and Litho are then processed the same as a blank without a mounted Litho. Example: RSC, FOL, die cut, etc. The pre-printed Litho must be resistant to cracking at score lines when box is formed. Some Lithos contain a clear coating on top of colors which, depending on type of coating material, can crack when corrugated paper is stretched around corners, etc.

The use of Litho boxes is very popular when product uses visual appearance on shelf to enhance "point of sale". Significant progress in direct printing on corrugated board has caused Litho printed boxes to decline in popularity.

CUTTING DIES

There are two kinds of cutting dies, namely rotary and flat. Their basic costs are predicated on the amount of steel rule die used. Every cut line, score line outer edges, circumferences of holes, slots, or perforated scores requires steel rule to accomplish. On flat dies, the cost of steel rule is approx $1.00 per linear inch. Total inches becomes die cost. On rotary dies, the cost is approximately $1.50 per linear inch. Total inches consumed equals rotary die cost. Like the printing dies, box suppliers mark up cutting die costs as much as 100 percent. Again, upon payment of die charge, the die becomes your property. Again, this is a one-time charge. Repair or replacement is the responsibility of the supplier you paid. You can transfer the die to another box maker. As previously stated you will have to know the size of your die, rotary or flat, number of pieces out per hit, and drum size for mounting rotary die. If everything checks out, you can move the die to your new supplier. It is difficult to modify or change a current cutting die. If a change is necessary, a new die may be required. Flat and rotary dies can produce one piece or a number of pieces for each corrugated blank that it comes in contact with. As the number of pieces per die increases, the cost for the die increases. Common cut lines between adjoining pieces reduce the cost a little.

Note: Additional pieces of steel roll will be required to cut waste into smaller pieces. This applies to flat or rotary.

Question?????

Does your box supplier have insurance to cover the loss or damage of your printing and cutting dies in his possession?? If not, do you???

SHOCK AND VIBRATION

Shock, vibration, or both can cause damage to your product during transportation. The interior packaging components or material within the corrugated box is designed or selected to protect your product against damage or abrasion. Abrasion vibration of your product against the inner surface of the corrugated box or corrugated inserts can wear a painted surface down to bare metal. Coatings applied to the contacting surface of the corrugated or a plastic liner or bag can eliminate the abrasion problem. Coatings can be applied to the corrugated board at the corrugater. Applying coatings to your box blank is quantity sensitive. Corrugater may say no because quantity is too small to economically set up on corrugater.

Many coatings are available to meet your needs, namely:

1. Resist abrasion
2. Resist oil and grease
3. Non-slip
4. Stain resistant
5. Water resistant

Check with your supplier.

Note: Unprotected silver will rapidly tarnish in a corrugated box due to the presence of sulfur vapors entrapped during the making of the paper.

VIBRATION

Various characteristics of individual trailers, and roadways cause vibrations to be impacted upon your product in transit. Screws and bolts can loosen which in turn makes internal components highly prone to damage from shock. There are vibration test machines for purchase and testing laboratories available for performing vibration tests.

SHOCK

Shock is a force directed upon your product when it impacts another object or surface. This can be caused by one of the following:

a. Dropping the packaged item
b. Sliding around in a truck
c. Falling off other loaded products in the truck
d. Rough handling at a distribution center
e. Falling off a conveyor

f. Being impacted by another heavier packaged product on a floor to floor sliding ramp
g. Everyday exposure to shocks transmitted by smooth and rough road surfaces

WHAT IS SHOCK?

It is impacting a moving mass onto a surface and slowing it to a stop over a distance and time. A simple example: drop an item 30 inches onto a surface and cause it to gradually stop over another 30 inches. The shock is 1 g. Sitting in our favorite chair we all experience 1 g. Now drop it again from 30 inches but stop the item within 1 inch. The result is 30 g's of shock. Drop again from 30 inches but stop within 1/16 inch. The resulting shock is 480 g's. A "g" is the force of the product's weight acting upon itself. In our chair, our weight is acting upon us. At 2 g's we would feel like twice our weight was acting on us. 10 g's would be 10 times our weight in force. In evaluating a product's sensitivity to shock, the term fragility is used. A product with a low tolerance to shock would be considered very fragile (10 to 25 g's). A product capable of withstanding 100 g's would be less fragile.

Stacked loads on pallets can also be affected by shock even though the load is not dropped. Over the road trailers generally transmit 2 g's of shock into the contents of the truck. Potholes or sunken roadbeds adjacent to overpasses can result in spikes to 5 g's.

At 2 g's, the bottom layer of a stacked load reacts like the layers above weigh twice their total weight. At 5 g's the load above will act like it weighs 5 times its gross weight. If the bottom, and to a lesser degree the layers above, is not protected against a stack load of 5 times the actual weight, the bottom layer can, and probably will experience crushing and product damage. You are not in the truck during its journey and therefore you will not know what your product or load was subjected to. Ride recorder instruments are available to measure shocks in transit.

Corrugated suppliers, as a rule, have difficulty designing corrugated interiors to protect to a shock level. Molded foam can be designed to approximate a specified shock level. Regardless of source and/or material used TEST THE PACKAGED UNIT BEFORE PURCHASING. There are package test labs available, especially since the arrival of UN-packaging specs.

To absorb shock at impact, the product must have controlled movement. Molded and fabricated foam can control various movement distances by varying bearing areas and foam densities. You want a gradual slowing to a stop that foam, peanuts, and various cushioning wraps can provide within their limits. On the other hand, an empty box will provide a fast movement and a quick stop. Shredded paper looks great when packed. However, through vibration, the product cavity enlarges and enlarges to the extent it is like the empty box situation. A number of years ago, there was a formula for controlled product movement distance, namely:

Two x drop height (inches) divided by product fragility (g's) equals desired movement.

Example: Drop height is 30 inches; fragility of product is 20 g's.

2 x 30 = 60. 60 divided by 20 = 3 inches of stopping distance. Stopping distance should occur gradually over the 3 inches. Stopping distance is <u>NOT</u> cushioning thickness. Cushioning thickness is <u>ALWAYS</u> greater than stopping distance.

Fragility is the degree of susceptibility to damage caused by shock. The lower the fragility numbers the higher the potential of damage. Very few commercial products have a known fragility level. If it becomes critical, a test for fragility must be undertaken.

There is a phenomenon to shock at impact. The cushioning is compressed during impact (product movement) and at completion of absorbing shock wants to return as close as possible to its starting thickness. This return causes a reverse direction shock to be imparted into the product. This shock is less than the original. High-speed cameras 400 fps has shown a couple hundred pound product being lifted up and off the cushioning inserts by the return shock. This aftershock should not be ignored if package vertical movement is restricted by a non-cushioning surface. Damage to your product during shipment can generate strong pressure to re-design the package. Do not rush to panic. Packaging costs money. Obsolete components of the old package is money wasted. New components may be more expensive. What caused the damage? Your product may have fallen off the tailgate of the truck because of poor handling. You cannot design that away and still make a profit. How many units have been similarly damaged over what time length? The answers to these and other questions should give you a clear course of action. If you are involved in hazardous products, shock and vibration testing is mandatory per D.O.T. and UN regulations.

PRICING YOUR BOX

Corrugated boxes and inserts are very quantity sensitive. Using a quantity of 1,000 as a nominal order, prices over 1,000 quantities go down gradually. Whereas order quantities under 1,000 go up sharply at an increasing rate. Example: a box that costs $.45 each in a quantity of 1,000 can cost $75.00 each in a quantity of five. Box prices are based on order/release quantity. A 12,000-blanket order with 1,000 released per month will be priced at the 1,000 quantities. The 12,000 quantities may have some affect on the profit margin but not the basic costs of material, labor, and overhead.

Material (corrugated board) is priced at cost per square foot. The corrugater establishes a price for a range of board feet. Example: 10,000 to 12,500 board feet will cost more per square foot than 12,501 to 15,000 square feet. The blank size, in square feet, required to make your box multiplied by the release quantity equals the board feet on your order. This quantity is compared to the various quantity price levels and your cost is determined. Different blank sizes, on a given order, cannot be combined to increase board footage and lower price. Each blank size stands alone. You may never see this price because it is given to your box maker. The box maker uses the board cost in determining the box cost to you. Increases in paper prices affect the board price directly. It will be discussed later. All components of the corrugated board are measured in pounds per 1,000 square feet (liners, glue, and medium). The medium is convoluted and as such a square foot of board surface requires a greater surface area of medium. The multiplying factors for medium are as follows, namely:

(A Flute) ------- 26# x 1.5 = 39.0#/1000 sq. ft.
(B Flute) ------- 26# x 1.3 = 33.8#/1000 sq. ft.
(C Flute) ------- 26# x 1.43 = 37.0#/1000 sq. ft.
(E Flute) ------- 26# x 1.27 = 33.0#/1000 sq. ft.
(F Flute) ------- 26# x 1.23 = 32.0#/1000 sq. ft.

Using 4# of glue for single wall and 8# for double wall, we now can compute the gross weight per 1,000 square of corrugated board in various flutes and combination of flutes, Example:

200 test c flute =

liners	=	42#	+	42#	=	84 pounds
medium	=	26#	x	1.43	=	37 pounds
glue	=	4#			=	4 pounds
				total	=	125 pounds

Note: different types of glues may have different weights per 1,000 sq. ft.

Approximate board weights:

125 C ------- 94#/1000 sq. ft.

150 C ------- 108#/1000 sq. ft.

150 B ------- 104#/1000 sq. ft.

150 E ------- 103#/1000 sq. ft.

175 E ------- 114#/1000 sq. ft.

175 Cc ------ 118#/1000 sq. ft.

175 B ------- 114.8#/1000 sq. ft.

200 C ------- 125#/1000 sq. ft.

200 B ------- 122.8#/1000 sq. ft.

200 E ------- 122#/1000 sq. ft.

275 C ------- 180#/1000 sq. ft.

350 C ------- 222#/1000 sq. ft.

Double wall weights:

275 CB --------------- 196#/1000 sq. ft.

350 CB ----------------- 205#/1000 sq. ft.

500 CB ----------------- 301#/1000 sq. ft.

600 CB ----------------- 349#/1000 sq. ft.

Note: The combination of (B) and (C) flutes in double wall board is the most common. (See generalized cost vs. quantity curve.)

TESTING FOR BOARD COMPLIANCE

Some corrugated sources will "low ball" a price to get the order. To offset the low price, they can use one or more of the following:

a. Lower test material used but print the certificate for the proper test value.
b. Reduce the quantity per bundle.
c. Reduce the number of bundles per pallet load.
d. Just invoice a quantity that can't be loaded on a full trailer.

There are three 200-pound tests being quoted; namely, using standard 42-pound liners, cheating -- using 38-pound liners, or new edge crush test (ECT) using 35-pound liners. You will not usually know which of the above is used in your box. Reputable sources will tell you and the others will not. Sheet plants and brokers do not make the board but they should know the reputation of their suppliers. Remember, the manufacturer's certificate on the box flap is no guarantee. There is a simple way to check the material in your box or insert. Cut a reasonably accurate 4" x 8" sample from the box or insert. Sample is to be void of holes or slots. Sample can contain score lines and printing. Weigh sample on a gram scale. The weight in grams will be same numbers as weight per 1000 square feet except decimal point on square feet value is moved one point to the left.

Example: 200 pound test weight/1000 square feet equals combination of weight of two liners (42# each) + expanded weight of "c" flute medium (37#) plus 4# of glue to **total 125#.** A 4" x 8" sample will weigh **12.5 grams** on a gram scale. To check your material, place your sample (4" x 8") on a gram scale and compare with weight values in pricing section. Remember to move decimal point one unit to the left. (125 pounds = 12.5 grams.)

The sample was not "climatized". It is close enough to have a very serious talk with your supplier. Some sources may use reprocessed Kraft paper in their linerboards. To maintain paper strength, heavier weight paper must be used. Example: to meet strength of 42-pound paper, 55-pound reprocessed paper is used. Advances in paper technology could change these weights.

Note: Do not forget to balance out your gram scale with zero weight before weighing samples.

For legal evaluation, send sample or samples to a test lab with testing capability. They will separate liners and medium, climatize components, weigh each, and present an official report. This is not cheap.

CONVERSION TABLE

POUNDS PER 100 SQUARE FEET
TO
GRAMS PER 4 INCH X 8 INCH SAMPLE

TEST VALUE	LBS./M SQ. FT.	GRAMS / SAMPLE
125 A	91.0	9.1
125 B	90.0	9.0
125 C	93.0	9.3
ECT 29 A	104	10.4
ECT 29 B	99	9.9
ECT 29 C	102	10.2
ECT 32 A	113	11.3
ECT 32 B	109	10.9
ECT 32 C	111	11.1
150 A	113.0	11.3
150 B	108.0	10.8
150 C	111.0	11.1
175 A	119.0	11.9
175 B	114.0	11.4
175 C	117.0	11.7
200 A	127.0	12.7
200 B	122.0	12.2
200 C	125.0	12.5
275 A	154.0	15.4
275 B	149.0	14.9
275 C	152.	15.2

350 A	**223.0**	**22.3**
350 B	**218.0**	**21.8**
350 C	**221.0**	**22.1**
ECT 48 CB	**175**	**17.5**
275 CB	**196.0**	**19.6**
350 CB	**205.0**	**20.5**
500 CB	**301.0**	**30.1**
600 CB	**349.0**	**34.9**

NOTE: ACTUAL SAMPLE SHOULD BE 4 INCH X 7-15/16 INCH. EXTRA 1/16 INCH ADDED TO MAKE SURE. IN AS MUCH AS THE 4 X 8 SAMPLE HAS NOT BEEN CLIMATIZED (SUBJECTED TO A SPECIFIC TEMPERATURE AND HUMIDITY) THE RESULTS CAN NOT BE USED IN A COURT OF LAW. THEY ARE VERY CLOSE AND CAN BE THE BASIS FOR A SERIOUS TALK WITH YOUR SUPPLIER.

FORMULA IS RATIO OF 1 GRAM TO 453.6 GRAMS AND RATIO OF SAMPLE SQUARE FEET TO 1000 SQUARE FEET.

PRICE CHANGES

Reduction in the price of a box or insert you are using usually does not happen unless your supplier is very conscientious or you apply pressure to your supplier. It is important to stay informed on current and past $$$$/ton of linerboard and medium paper prices. The cost/ton is reported monthly in purchasing magazines. As the cost of paper changes, the cost of corrugated board changes. Do not forget, paper is only a part of the cost picture. Labor, overhead, commissions, and profit are in the price you pay but are not affected by fluctuations in paper prices.

PRICE INCREASES

Price increases are usually generated by the paper mills but not always. Paper stock in large rolls is sold to foreign countries. The paper companies reportedly get a good price per ton. They have been trying for years to get the domestic price per ton close to the export price. The domestic market, through competition, fights to get lower prices. The result is an up and down cycle of paper prices. Again, paper is only a part of the selling price of a box or insert. A 10 percent increase in paper does not mean a 10 percent increase in your box price. Did labor, overhead, commission and utilities go up 10 percent? Should a 10 percent increase in profit be accepted? There are increases in non-paper areas that can be substantiated but not in every paper increase. The real effect of paper increases on price is a factor of release quantity. In small quantities (100, 200, 300), labor and set-up account for much of the selling price with material being a small percentage. In a 1,000 quantity order, material can account for 65 percent of the selling price. Higher quantities can increase the percentage paper consumes in the selling price. Using the 65 percent figure, a 10 percent increase in paper should affect box price no more than 6-1/2 percent in a quantity of 1000. These numbers can change, but never pay the full paper percent increase on box pricing.

ORDERING CORRUGATED BOXES OR INSERTS

There are three ways to order boxes or inserts, namely:

1. Dimensioned drawing of item.
2. Sample of item.
3. Verbal description of item.

All manufactured corrugated items have a plus or minus tolerance in their dimensions. On boxes, the accepted tolerance is plus or minus 1/8 inch. This means a 10-inch dimension can range from 9 7/8 inch to 10 1/8 inch. Let us review the three ways of ordering.

DIMENSIONED DRAWING

This is by far the best way to order your box or insert. If the inside dimensions of the box provide for the 1/8-inch tolerance, the box manufacturer should deliver a satisfactory container. Likewise, the insert should fit into the box. Changes can be noted on the drawing along with date of change. A basic corrugated box drawing, with no printing, can consist of a verbal written description with inside (L x W x D) dimensions listed on the drawing along with all other pertinent data such as test, flute size, desired manufacturers joint (glue, stitch or tape) type of box (RSC, FOL, HSC, telescope etc.) A dimensioned drawing showing all four length and width panels laid out flat clearly shows the length, width and depth dimensions of the box. On a drawing with written specifications, be very sure to define the length, width and depth dimensions with the letters L, W, H or the actual words. Remember, the box opening is defined by L and W dimensions. If your supplier has a record of the correct specifications, you need only reference the item's part number or description along with the quantity to be run to order.

SAMPLE OF ITEM

This can work but it is subject to major problems. Remember, the tolerance. Was your sample exactly to the required dimension? Or was it a little smaller or larger? If the new order is run on the minus side of a sample already on the minus side, you can end up buying an item that is too small for your product. Neither you nor your supplier knows the intended dimension of the item. Who is at fault? If your supplier still has the actual sample, he could prove he is within tolerance. Is it the actual sample? Is it a box from the current run? This approach has problems. The use of samples is a popular way of providing specifications to a new or potential supplier. Slight dimension variances will not affect the potential supplier's price but, as said before, it can affect the fit of your product to the new suppliers box dimensions. If you must use a sample, identify it with your signature or some other marking.

VERBAL DESCRIPTION

If done properly, this can also work. It also has potential of serious problems. Did you do the following?

1. Provide inside dimensions.
2. Dimensions in proper sequence (L x W x D).
3. Style of box (RSC, FOL, HSC, etc.).
4. Which manufacturers joint (glue, stitch, tape).
5. Test value of material and flute size.
6. Printed or plain.
7. Quantity.
8. Surface color (Kraft, Oyster white, or solid white).
9. Verification that supplier thoroughly understands your requirements.

If your supplier makes an error writing down your specifications, you will not receive what you want. Can you prove what you said verbally in person or over the phone? This can result in a nasty relationship with your supplier. Who pays for the mistake? Protect yourself, write the specifications down and fax a copy to supplier. Have supplier verify receipt of fax.

Note: Items one through eight above, when typed or written, along with a part number, can be considered a dimensioned drawing for a corrugated box. Corrugated inserts are usually more complicated and as such will require a dimensioned drawing.

As most purchasing people know, your order and your supplier's confirmation constitute a contract. Your company's bottom line is involved. Make sure your part of the contract is correct in all aspects.

Note: The national average for delivered vs. order quantities is plus or minus 10 percent. Quantities above 10 percent overage can be returned. Some suppliers overrun significantly and deliberately. I have seen 100 percent overruns and more. If overrun is excessive and usable, tell your supplier to re-invoice at a lower price/1000 commensurate with degree of overrun.

Overrun limits of 10 percent is difficult on quantities under 500. The difficulty increases as the order quantity goes down. Discuss this with your supplier before ordering.

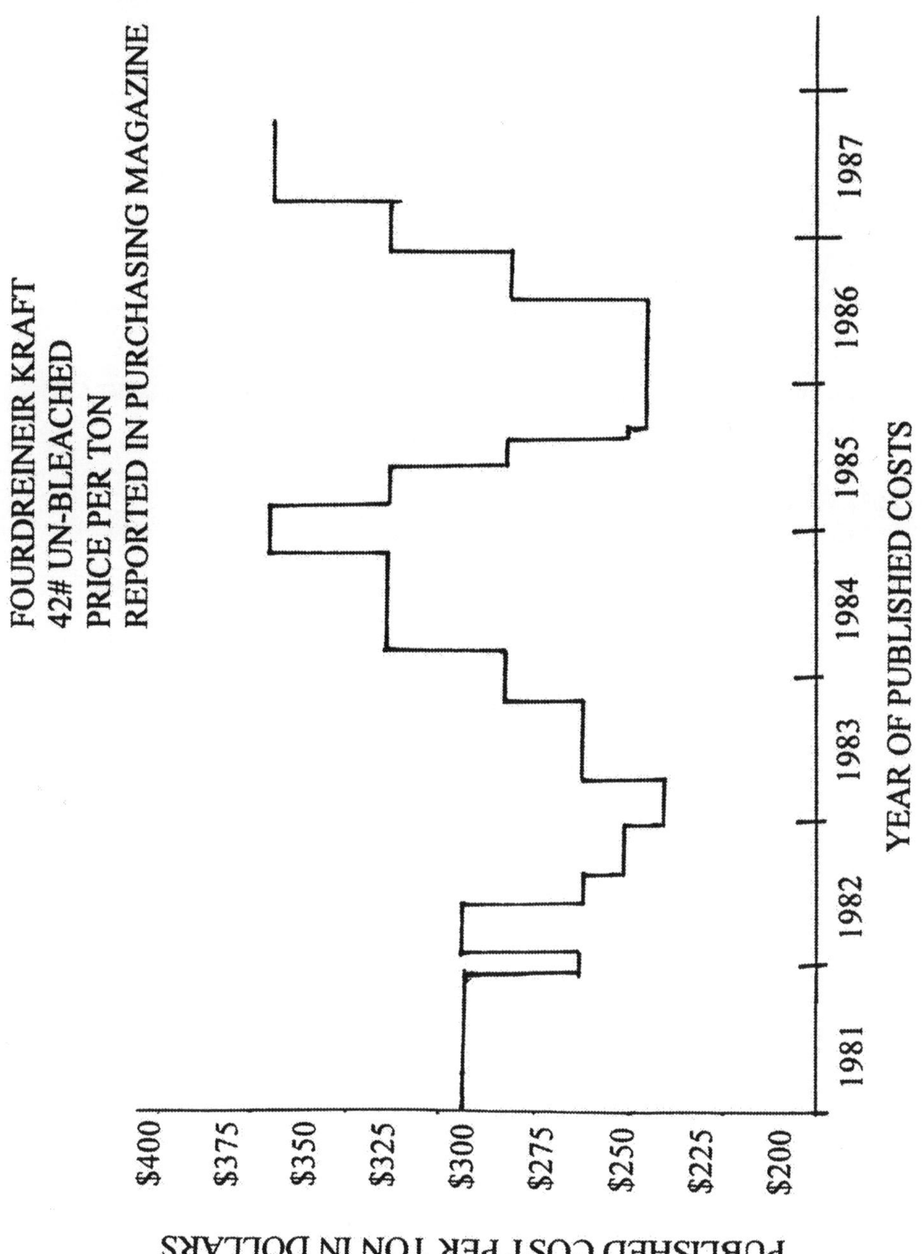
FOURDREINEIR KRAFT
42# UN-BLEACHED
PRICE PER TON
REPORTED IN PURCHASING MAGAZINE
PUBLISHED COST PER TON IN DOLLARS
$400
$375
$350
$325
$300
$275
$250
$225
$200
1981
1982
1983
1984
1985
1986
1987
YEAR OF PUBLISHED COSTS

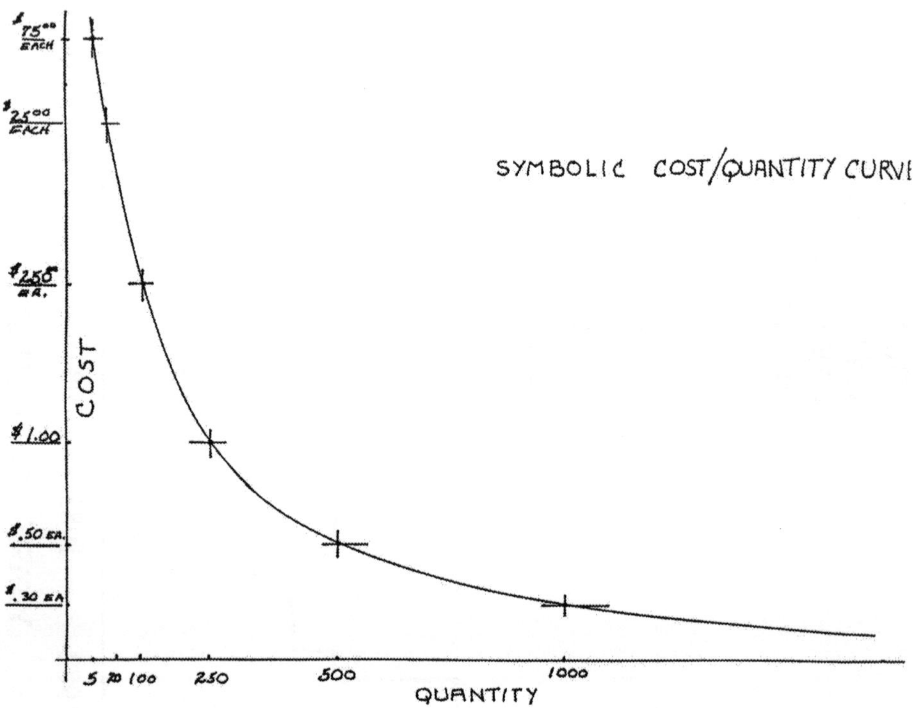
$75.00 EACH
$25.00 EACH
$2.50 EA.
COST
$1.00
$.50 EA.
$.30 EA
5 20 100 250 500 1000
QUANTITY
SYMBOLIC COST/QUANTITY CURV

CHAPTER 2

INTERIOR PACKAGING

●●

IN THIS CHAPTER

●●

●●

INTERIOR PACKAGING

Expanded Polystyrafoam (E.P.S.)

E.P.S. is a small (like sugar) beaded material that expands when subjected to steam. It can be expanded into a product by a molder or expanded into a large loaf to be cut up into sheets, pads, or product. Molded parts give you repeat dimensional stability, variety of densities, reasonable piece price, and high tooling costs. Sources for molded parts are not plentiful. With lack of competition in some areas of the country, prices and tooling costs can be suspect.

There are a number of E.P.S. fabricators, and there are three types of E.P.S. suppliers, namely:

1. Makes large loaf (10 ft. x 8 ft. x 4 ft.) and cuts as needed to make sheets, pads, product. Also sells to other fabricators.
2. Buys block or sheets and fabricates sheets, pads and parts.
3. Makes nothing. Buys and sells from (1) or (2).

Fabricators usually limit the available density to one pound. Some may get into two-pound density.

Note: Density is weight per cubic foot of material.

The tolerance on densities is 10 percent.

WORKING LOAD VALUES FOR E.P.S.

1 pound	3	pounds per square inch bearing surface.
2 pound	11	" " " " " "

TOOLING COSTS E.P.S.

Tooling cost for fabricated parts with holes, cut outs, and odd shapes can run $50.00 to $300.00. Molded part dies can run $5000 to $15000. This varies to size of part or number of product parts in mold.

FREIGHT COSTS

Fabricators are usually local which results in lower freight charges. Molders may be 300 to 600 or more miles away. At a class 300 or more, freight can be significant.

FOAM MATERIALS

There are various foam materials on the market. Some of them are listed below:

1. Polyethafoam: This material is fabricated from molded planks, which contains a curved skin on its outer surface. The skin curvature usually has to be removed before fabricating your parts. Plank widths can be 24 inches up to 48 inches. Densities can equal six pounds per cubic foot. This material is not cheap in one-pound density requirements, but may be your only choice in heavy product loads.

2. Expanded Polystyrene: This was previously covered. Densities of this foam have been equated to static load support.

1 pound density	3 pounds per sq. inch.
2 pound density	11 pounds per sq. inch.

Most fabrications use 1-pound densities, with a few in 2-pound densities. Some government contracts call for 3 pound density parts. Molded parts can be made in many densities. Remember, density can vary plus or minus 10 percent.

3. Foam in place: This consists of resin and a catalyst which when combined creates expanding foam. A special nozzle is required to dispense the foam. The more material released through the nozzle, the more foam is produced. Atmospheric conditions can affect the expansion of the foam. The expanded foam will adhere to all surfaces. To keep this from

happening, product and/or any other surface must be separated from foam with paper or plastic separator material. Excess rise of foam in corrugated box must be trimmed off before closing container. Densities are usually low. Pre-molded cavities can be made with this material to your product and box specifications.

4. Foams are also available in the following:

    a. Polyurethane
    b. Polyethylene
    c. R-cell
    d. Others

Note: all molders or fabricators do not handle all foams. Very few if any suppliers of foam will "knock on your door". You will have to use word of mouth or "let your fingers do the walking". Also, not all foams can be molded.

HONEYCOMB

Honeycomb is a manufactured paper-packaging product. It contains up to three paper components like corrugated, but the material between the linerboards is significantly different. The paper used is the same as used in corrugated boxes. Again, the liner paper can range from 26 pound per 1000 square feet on up to limits controlled by honeycomb source. The paper used to make the middle or core of the honeycomb has the same range of paper weights. The liner papers, the core paper and the cell size can be varied to accommodate your wishes in pounds per square inch load factors and cost. The width of the glue line within the cells can affect the load bearing strength of the honeycomb part. Honeycomb has a grain direction and can affect strip stability when the strip is partially pre-cut into segments within its length. Availability of honeycomb is limited to a few locations within the USA. New companies or mergers may change this. Honeycomb is very light in weight and as such carries a significant freight rate. Small quantities may be too costly to use.

BUBBLE WRAP

Bubble wrap is a series of plastic bubbles in line on a plastic backing sheet. The commercial grade is basically polyethylene. Military grade is available.

Bubble wrap can be purchased in rolls or sheets. The basic roll width (generally 48 inches) can be cut down to lesser widths to meet your needs. Check with your shipping supply source for available or possible roll widths. Roll width and quantity of rolls can result in unusable widths (scrap), that you will pay for whether it is used by you or not

Sizes of bubble wrap are as follows:

3/16 inch
5/16 inch
½ inch

The above sizes reflect the height of the bubbles only. Application of bubble wrap is with bubbles facing toward product being protected.

APPLICATIONS

Small (3/16) bubble:

Small, light weight, and/or multiple items. Surface protection.

Medium (5/16) bubble:

Surface protection, cushioning, and void filler

Large (1/2) bubble
Void filler

Features of bubble wrap:

- Transparent
- Lint and dust free
- Non-abrasive
- Non-absorbent
- Lightweight
- Flexible
- Non-corrosive
- Non-toxic
- Shock absorbing potential

Note: To absorb shock, the product must move in a controlled manner until it comes to a stop. Insufficient wraps around a product can cause said product to prematurely come to a stop and increase the shock transferred to the product. Drop test package whenever possible before making packaging design official.

BULK FILL MATERIAL

Bulk fill material comes in various shapes. It is generally expanded polystyrafoam (EPS). It usually comes in 14 or 20 cubic foot bags. This is a good cushioning material for many products with a medium or above fragility levels. However, it can produce dust and static electricity.

To keep your product from shifting around in the box, overfill the box a minimum of 1/2". This forces the bulk material to interlock when box is closed. Otherwise the vibrations experienced in transit will cause your product to move to the inner surface of the box and be seriously exposed to damage. The distance between your product or products and the inner surface of the box established at the time of packaging will be lost.

PAPERBOARD – CHIPBOARD

Paperboard or chipboard is a solid fiber material. It can be purchased in a variety of thicknesses. The thickness under 1/8" is defined as "points of thickness". Example: 30 point is decimally .030 inches. Point values of 20 to 120 are the ranges usually available. Chipboard is usually gray in color. This is due to the black ink in newspapers, which is the primary source of raw material. The hard backing on a pad of paper is chipboard. Chipboard has a grain direction. When ordering pads, sheets, or partitions, the grain direction can be very important. In the past and possibly still today, chipboard or paperboard consisted of 10 separate layers of material combined into one regardless of final thickness. This will permit a white surface on one or two surfaces. This, of course, is obtained by using bleached material instead of newsprint in the first or last of the 10 layers. In years past, water for the chipboard process was not purified and contained some contaminants. Due to this process, chipboard was not permitted to come in contact with food products. Chipboard has excellent anti-abrasive qualities.

HOMOSOTE

Homosote is basically the same as paperboard. It can be procured in significantly greater thickness. It can be die cut. Homosote is not a cushioning material for light to medium weight products. It is good for positioning a heavy product and keeping it there.

PARTITIONS

Partitions, in packaging, are a quantity of cells adjacent to each other. They can be made from corrugated or chipboard. They can be made with cell sizes ranging from ½" and up. They can be sold as component pieces or pre-assembled. As pieces, you have to put them together. As pre-assembled, they are collapsed to reduce space and packed for shipment. Upon receipt, you open them up and place them in your box. Partitions are used to separate fragile parts such as glass items or to prevent abrasion or both. Remember, corrugated and chipboard has a grain direction. The material can flex parallel to the grain direction and resist flexing perpendicular to it.

Air cells (not containing parts) can be provided on outer sides of partitions. This provides a "set-back" of the part cells from the inner surfaces of the box sides for special needs. Air cells (no parts) can be provided internally between parts. It is suggested that overall length and width of partition be 1/8" less than the length and width of the corrugated box. Delivery quantities can range +/- 10 percent of ordering quantity.

CHAPTER 3

WIREBOUND CONTAINERS

••

IN THIS CHAPTER

••

••

WIREBOUND CONTAINERS

Wirebound containers are wood containers that contain multiple rows of heavy gauge wire to hold the four independent sides together. Wirebound containers (boxes) consist of three components, namely:

a. Bottom or pallet base
b. Four joined sides
c. Top cover (when required)

All three components can interlock. This means no strapping (steel or plastic) or stretch wrap is required for closure.

FEATURES

The four sides come collapsed to save space with bottom (pallets) and tops unitized separately. The four sides are opened up and placed around interlocking pallet or base. The interlocking top is placed within four sides.

The wood used as vertical members on sides and top can range from 1/10 inch up to 1/2 inch in thicknesses. Mixed thicknesses can be combined to achieve strength where required and control costs.

The coverage of wood sides can range from 3 four-inch wide slats up to full coverage with plywood. To provide interlock on bottom, dimension stability at top, and resistance to bowing of vertical members under compression, cleats are used. “Cleats” are wood members approximately ¾” x ¾” or ¾” x 3”. The top or bottom cleats can be all ¾” x ¾” or a combination of ¾” x ¾” and ¾” x 3” as required by box design. The anti-deflection cleat is placed half way up the sides and can be referred to as the “belly band cleat”. This is usally ¾” x ¾”.

To close the open corner of the four side panels, a “salley” tool is the proper tool. A good heavy-duty screwdriver can also work. There are loops at the end of wires. One end has small loops and the other end has large loops. The small loop goes through the large loop and is forced back 180 degrees to pull sides into a tight closure. To open completely enclosed box, reverse loop closure process. Remove lid and side panels to gain access to contained product. Wirebound containers have become less popular over the last decade but still have good attributes especially on heavy and irregular-shaped products.

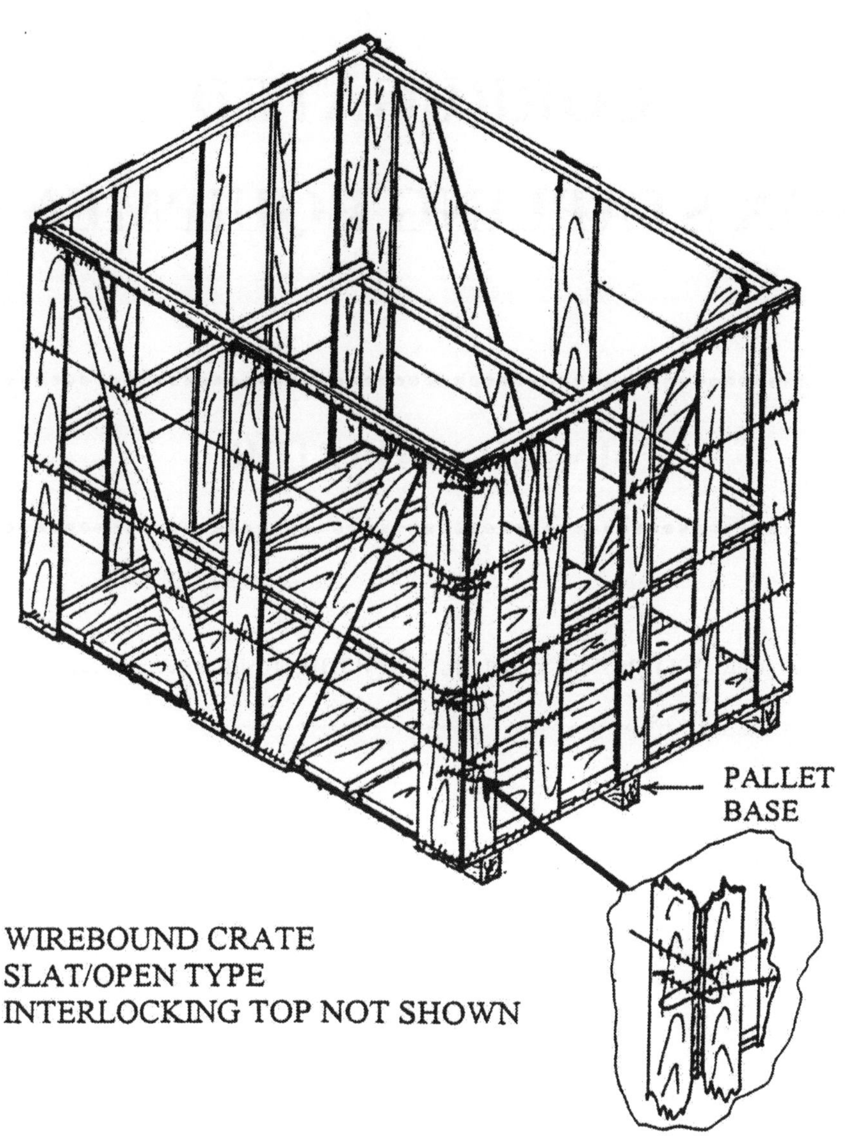
PALLET
BASE
WIREBOUND CRATE
SLAT/OPEN TYPE
INTERLOCKING TOP NOT SHOWN

CHAPTER 4

CORRUGATED BOX STAPLING EQUIPMENT

••

IN THIS CHAPTER

••

••

CORRUGATED BOX STAPLING EQUIPMENT

There are various types of carton stapling equipment available for purchase, specifically:

1. Pedestal wire-fed staplers
2. Small 5/8 inch crown staple units with attached bayonet anvil
3. Manual or pneumatic 40 staple units
4. Pneumatic 1000 staple units

The last two units are the most popular for sealing closed a filled corrugated box. The first unit is usually used to seal the bottom of an empty corrugated box. The stapler with the bayonet is preferred when stapling corners of corrugated trays and stapling flaps or cartons when the penetration of the stapler anvils would cause damage to the packaged product.

The quality of the box closure is 100 percent dependent upon the quality of the formed staple. The anvils in the stapler that form the staple are sensitive to abuse and can go out of alignment. Misaligned anvils will cause zero or only partial forming of staple. Internal pressure on flaps, due to weight of contents, can cause improperly formed staples to fail and contents of box to fall out.

Free triggering a 40 staple unit, in the past, has not shown the quality of the formed staple when applied to a corrugated box. When triggered into the air, the anvils are not supported as they are when applied to a carton. Conversely the 1000 staple unit, during that same period, duplicated the carton staple quality when free triggered. The bottom line is to frequently check your staple closures on your corrugated boxes and repair or replace defective anvils.

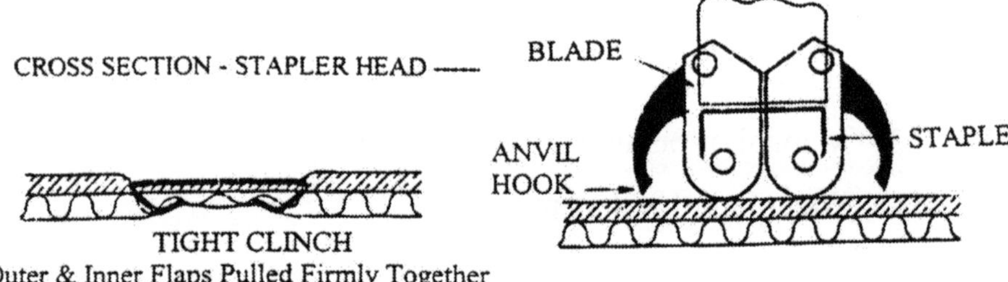

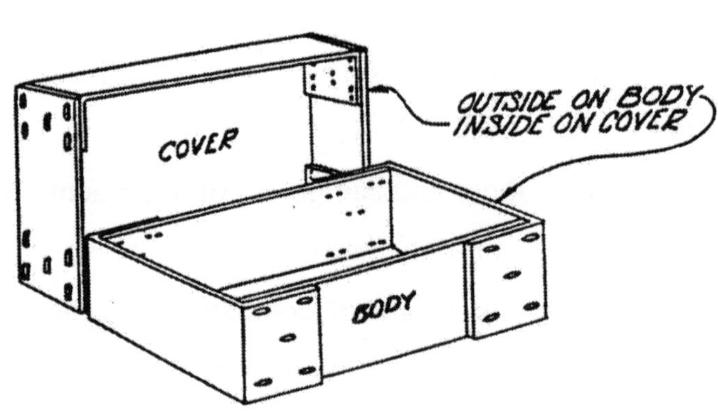

CHAPTER 5

TAPE AND TAPING EQUIPMENT

••

IN THIS CHAPTER

••

••

TAPE AND TAPING EQUIPMENT

WATER ACTIVATED CARTON SEALING TAPE

PAPER TAPE

Paper tape comes in multiple basis weights (pounds per 1000 square feet) depending on light, medium or heavy-duty applications. Tape comes in roll widths up to 3 inches wide. Tape can be purchased in natural color or white. Tape can be activated by a wet sponge or a tape dispenser with a water reservoir.

REINFORCED PAPER TAPE

Reinforced paper tape contains fiberglass filaments in a variety of patterns sandwiched between two layers of kraft paper. Tape comes in roll widths up to four inches. Different paperweights are used for regular or heavy-duty applications. Tape can be purchased in metric roll widths.

PRESSURE SENSITIVE CARTON SEALING TAPE

Pressure sensitive tape primarily comes in two (48mm) inch widths. For a hand held dispenser, tape comes in 55 yard and 110 yard roll lengths. One 110 yard roll cost less than two 55 yard rolls. Machine applied rolls measure approximately 14 ½ inches in diameter. Pressure sensitive carton sealing tape comes in a variety of variables namely:

1. Film thickness
2. Adhesive thickness
3. Type of adhesive
4. Manufacturing process to make film (blown or cast)

All manufactures do not provide all variables.

Example: Variables in adhesive include the following:

1. Acrylic
2. Hot melt
3. Natural rubber

There are an estimated seven manufacturers of tape. Many specialize in one adhesive. Of the seven, shipping supply companies probably limit the manufacturers to three or less and tightly control the variables they keep in stock.

It all comes down to what does the job at the most economical price.

See next illustration on next page for closure techniques using paper or reinforced tape. These same techniques can be considered when using pressure sensitive tapes.

REINFORCED TAPE

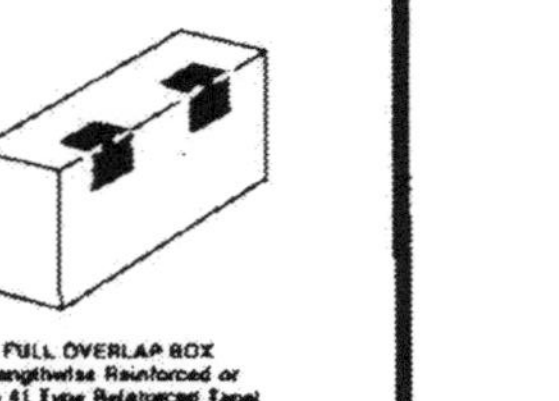

1. Apply one strip of tape to far flap (2½" overlap on each end)
2. Close flaps and seal center seam.
3. Fold down and press overlaps on each end.

Closure techniques

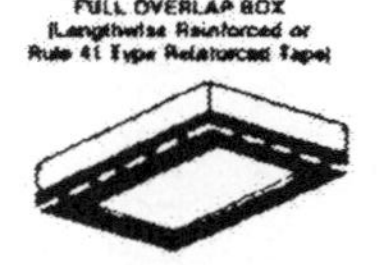

PARTIAL OVERLAP BOX (Rule 41 Type Reinforced Tape)

REGULAR SLOTTED CONTAINER (Rule 41 Type Reinforced Tape)

FULL OVERLAP BOX (Lengthwise Reinforced or Rule 41 Type Reinforced Tape)

FULL OVERLAP (Lengthwise Reinforced or Rule 41 Type Reinforced Tape)

TELESCOPE BOX (Lengthwise Reinforced or Rule 41 Type Reinforced Tape)

TELESCOPE BOX (Crosswise Reinforced or Rule 41 Type Reinforced Tape)

SLIDE STYLE BOX (Rule 41 Type Reinforced Tape)

1-, 2- and 3-PIECE FOLDER (Rule 41 Type Reinforced Tape)

5-PANEL FOLDER (Rule 41 Type Reinforced Tape)

PAPER TAPE

1. Apply first strip to far flap, then close flaps and seal center seam.
2. Press down overlap on ends (2½" overlap on each end).
3. Seal edge seams with 3-inch overlap beyond corners.
4. Bend around sides and pull tightly into position
5. Fold corners over the top and press firmly, then fold and press top.
6. Repeat 3, 4 and 5 on the other end of box

Closure techniques

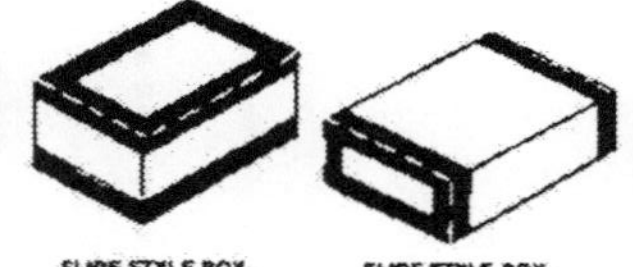

1-, 2- and 3-PIECE FOLDER

SLIDE STYLE BOX

SLIDE STYLE BOX

5-PANEL FOLDER

CHAPTER 6

PALLETS

●●

IN THIS CHAPTER

●●

●●

PALLETS

There are, as most people know, various styles of pallets, namely: (See attached illustrations.)

a. Two way entry
b. Four way entry
c. Single face
d. Double face
e. Block
f. Mil spec.
g. With wings
h. Flush
i. Expendable
j. Returnable
k. Hardwood
l. Softwood
m. Heavy duty

The style of pallet you want to use is your option. Let us discuss the material specifications you are going to use in making the pallet.

DIMENSIONS

Deck boards and runners are cut to your length specifications. There are standard lengths called "cuttings" that are used to determine material costs. Cuttings are usually 24 inch, 30 inch, 36 inch, 40 inch, 42 inch, and 48 inch. There are values above and below these numbers. If you specify a board length 1/8 inch over a cutting number, you are charged the next highest cutting length. Example: 43-inch runners will be charged out as 48 inch; 31-inch length deck boards will be charged out as 36-inch. The longer chargeable lengths increase the number of board feet used to make the pallet and in turn increase its cost. One board foot is a nominal 12" x 12" x 1". All components are calculated in nominal board feet. Nominal one-inch thickness has dropped in real thickness from 7/8 inch of many years ago to approximately 11/16 inch today. It is still charged as one inch. Normal wood dimensions can be referred to in two ways:

1. Actual inches – 2", 1½", 1¼", 1"
2. Multiples of ¼" – 8/4", 6/4", 5/4", 4/4"

Either way, these are nominal and not finished dimensions.

Nominal 2 x 4 inches in softwood is really 1-1/2 x 3-1/2 inches. Hardwood may be a little larger. Again, you are charged the nominal value. Nominal 6-inch and 4-inch wide deck boards vary in actual width but never exceed the nominal width. Deck board thickness of 5/8 inch full and 1/2 inch full can be obtained by re-sawing other board sizes purchased for this purpose. 2" or 8/4" nominal runners can be reduced to 6/4 or 5/4 nominal widths to reduce cost. The actual thickness of the 6/4 runner drops to 1-3/8 inches or 1-1/4 inches. 5/4 runners drop to 1-1/8 inches or 1 inch in width. 6/4 runners contain 25 percent less board footage than 8/4 (2 x 4) runners. On single face pallets, the thinner 6/4 runners are susceptible to "rolling over". To counteract this, bottom boards may be added (double face). Another choice is to use 11 gauge hardened screw nails approximately 3 inches in length to stop roll over by the runners.

DECK BOARDS

Leading edge and trailing edge of pallet should be nominal 6-inch wide deck boards with three nails per intersection with runners. This prevents "racking" of the pallet when hit by the fork truck. (Racking is when a pallet is forced out of square.) Four-inch wide boards require two nails per runner intersection. Interior deck boards can be 4-inch nominal. The number of deck boards and the clearance between boards depends on weight of load and size of material transported. Many pallet suppliers across the country have a computer program that can analyze your pallet for strength, roll over and compatibility with your storage racks. With the environmental considerations, lumber will continue to be in demand for housing and pallets. This will have a bearing on lumber prices and the cost of your pallet. Look at your specifications (if you have any) and look for ways to reduce the board footage. Do you ship the pallet with your product and never see it again? Do you get it back and reuse it? How much pallet do you need to support your product during shipment? Knowledge of wood material, and a good supplier providing answers to the above questions should help you to arrive at the optimum solution. "Winged" pallets contain deck boards that extend beyond the outer sides of the runners. Wings can be any dimension 1/2 inch or greater. The wing can provide a "grabbing" surface for stretch wrap, which in turn can hold the load down onto the pallet.

PALLET DIMENSIONING

Basic "runner or stringer" pallets are dimensioned with the runner length first and the width (deck board length) second. A 48" x 40" pallet is significantly different from a 40" x 48" pallet. See pallet sketches.

FOUR WAY ENTRIES

To increase a two-way entry pallet to a four-way entry pallet requires the notching of the pallet runners. These notches can be square cut or cut with radiuses. Care must be taken to insure that the remaining portion of the runners (vertical thickness above notch) can support the load if said load is placed up onto a storage rack with only front and rear support members.

NAILS

"Blunt" nails will reduce the splitting of deck boards. Chisel point nails will increase the splitting of the deck boards. Splitting becomes more of a problem the closer the nails are to the ends of the deck board. Nails should be "hardened". They come in various styles and gauges. Hardened screw nails in 11 gauge are preferred for single face pallets, especially when 6/4 runners are used. Small gauge/soft nails on single face pallets can permit runners to roll over and pallet to collapse. Runners (2x4) with soft nails have been rolled over BY HAND. Hard 11 gauge nails have significantly reduced this problem. Double face pallets do not require hard 11 gauge nails. It comes down to economics. Proper nails/single face vs. other nails/double face. In the past, a National Wood Products Association member with a Doctorate Degree has recommended hardened 11 gauge screw nails on single face pallets.

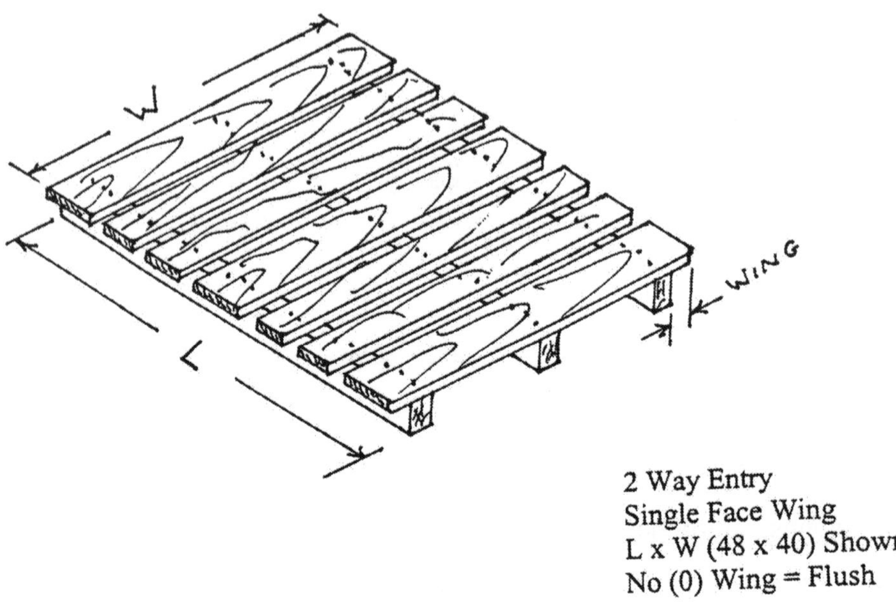

2 Way Entry
Single Face Wing
L x W (48 x 40) Shown
No (0) Wing = Flush

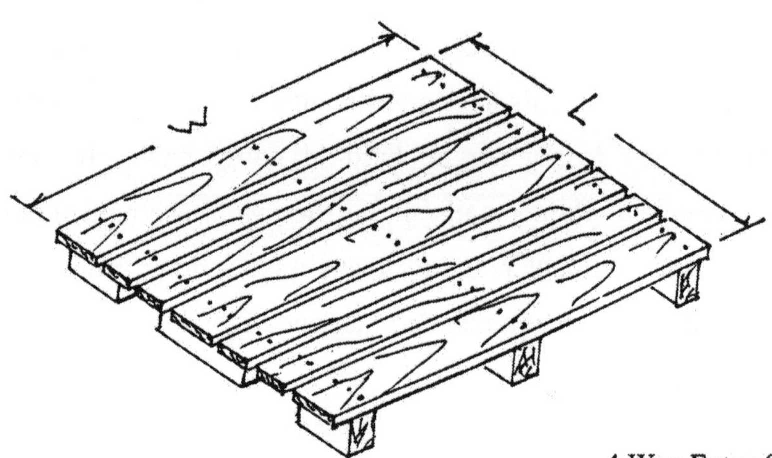

4 Way Entry (Notched)
Single Face Wing
L x W (40 x 48) Shown
No Wing = Flush

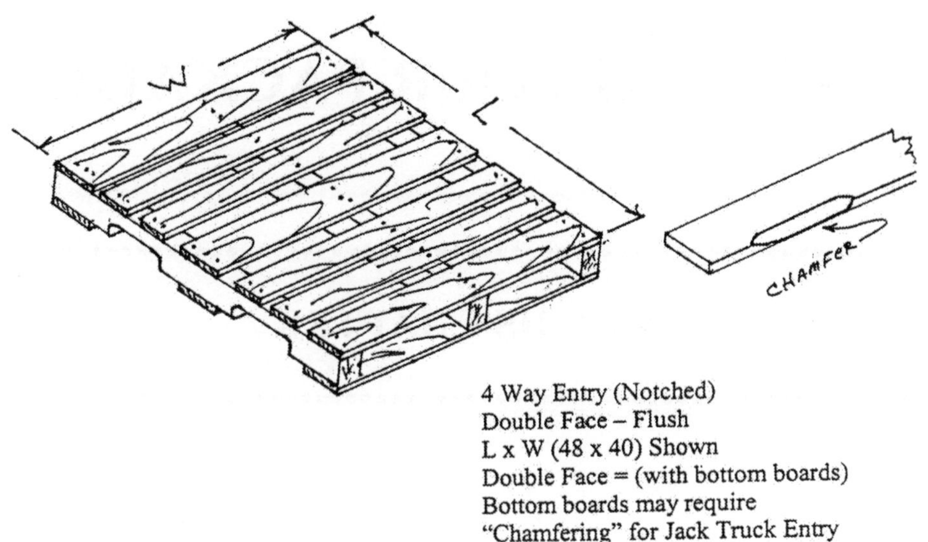

4 Way Entry (Notched)
Double Face – Flush
L x W (48 x 40) Shown
Double Face = (with bottom boards)
Bottom boards may require
"Chamfering" for Jack Truck Entry

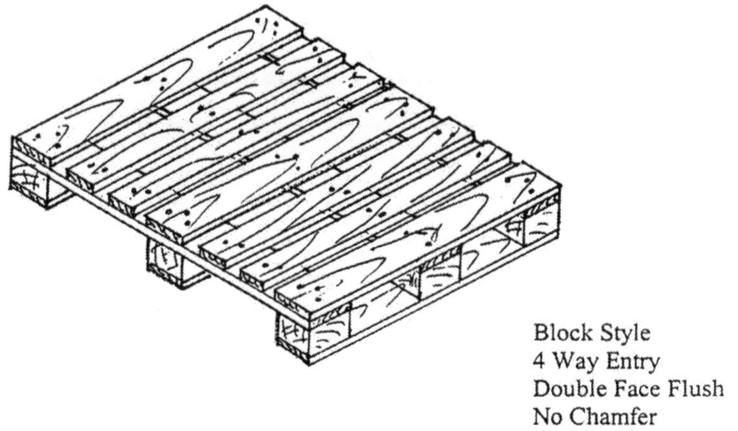

Block Style
4 Way Entry
Double Face Flush
No Chamfer

CHAPTER 7

TESTING YOUR PACKAGE

●●

IN THIS CHAPTER

●●

●●

TESTING YOUR PACKAGE

Before any packaging component is purchased, it should be tested to insure that it functions properly. This definitely applies to the overall package. There are test labs available to subject your packaged product to vibrations and impacts mirroring those encountered in transportation. For a fee or hourly charge, they will evaluate your packaged product and submit a report on its success or failure. A good report can help you recover any claims on a damaged product. You can also try to do it yourself.

IN-HOUSE TESTING

Vibration testing is a little difficult without a proper machine. You could drive your packaged product around your local area in a truck for a week. Drop testing is a little simpler depending on size and weight of packaged product. A national organization for many years advocated a 10-drop test as follows:

1. A corner of carton containing manufacturers joint.
2. One of three edges radiating from corner test #1.
3. Same as #2.
4. Same as #2.
5. One of six flat sides.

6. Two of six flat sides.
7. Three of six flat sides.
8. Four of six flat sides.
9. Five of six flat sides.
10. Last flat side.

Note: Six flat sides consist of top, bottom, two ends, and two sides.

The drop height will vary due to product size and weight (the lighter and smaller the package, the greater the drop height). Consider the normal position you would carry the package and its distance to the floor. This would be its drop height. Would you carry it waist high? Shoulder high? Or on a pallet?

If your product is undamaged after vibration and impact testing, you have a package that will survive normal handling. If you were to omit the vibration test, you would have a false sense of protection. Vibration loosens fasteners, which can make internal components weaker and in turn cause damage. Package insert components can lose some of their shock dampening features when subjected to vibration.

To be safe, if your product has a significant value and you cannot do both impact and vibration tests yourself, pay a reputable testing service to do it for you. This is extremely important for hazardous products or materials.

CHAPTER 8

STRETCH WRAPPING

IN THIS CHAPTER

STRETCH WRAPPING

Stretch wrap is a thin plastic film that can be stretched to a length significantly greater than its original length. This increased length can be wrapped around your palletized product and hold the load together until it reaches its destination. Properly stretched wrapped loads can eliminate the need for metallic or non-metallic banding plus various surface "strap protectors".

Stretch wrap film can be applied manually or by a variety of stretch wrap machines. It is the effect of the stretched film trying to return to its original length that holds the load together and to the pallet. The most popular machine applications are spiral wrapping. For example, a 20-inch wide roll of film and spiral wrap machine will turn and cover up to the top of your load and back down to the bottom of your load. Controls, within the machine, can vary the number of wraps, the amount of overlap per wrap, and the number of wraps around the pallet. The type of machine can also control the amount of stretch on your film. Depending on the machine, the amount of stretch can vary between 3 inches up to 30 inches on a 10-inch initial length. The additional film length, when stretched, has to come from somewhere. The extra length comes from reduction in film thickness and roll width. Reduction in roll width is called "neck down".

When stretch film is manually applied, the amount of stretch is controlled by how hard you pull the film around the load, which is stationary.

When stretch film is applied by machine, the amount of stretch is controlled by one of two ways:

1. The film roll is mounted on a spindle with an attached brake. The turn of the pallet load pulls the film against the "braked" roll, which causes the film to stretch. In the past, this approach produced small amounts of stretch like 3 inches on 10 inch original.
2. The film is mounted on a spindle without a brake. The film is passed through and over two special motorized surfaces, which operate at two different speeds. The difference in speeds causes the film to stretch. The speeds of these two surfaces can be varied to cause different amounts of stretch. This approach is called "pre-stretch". On machines with pre-stretch capabilities, the amount of stretch obtained is maintained to the pallet load by a control called "force to load". If the force to the load is not proper, the amount of stretch obtained in the pre-stretch will be reduced proportionately.

The more you stretch your film the less film is used to wrap your load of product. The less film the less material cost of the film. Film is sold by the pound. The film on a pallet load of product can easily be cut away and weighed. The weight times the cost per pound is material cost per load wrapped. Different film gauges and different amount of stretch can be tested. The cost "weight" of the film plus the change in physical protection of your product will determine what parameters are best for your company. Bring in a stretch machine and film specialist to verify your conclusions.

FILM AND FILM COST

There are many gauges of film. 100 gauge is one mill in thickness. 80 gauge is 8/10 of one mill thickness. The higher the gauge, the more the film weighs. The higher the original gauge, the higher the resultant gauge after stretching.

Film can be manufactured by either "cast" or "blown" process. At one time there were differences in stretch wrap produced as a result of these two processes, which may or may not still be in effect. It is highly possible that they are equivalent.

Film is priced on cost per pound of film. A roll of film consists of a "core" plus stretch film around the core. There should be a known weight of film per gauge on a roll. The full weight less the core weight equals the chargeable pounds of film. If you are charged by the roll, they should also tell you the number of pounds of film that should be in the roll. My suggestion is to insist on being charged by the pound of film but get information on the number of pounds per roll and the number of rolls. Trade journals will keep you up to date on film costs per pound trends that you can compare with your costs. Frequently check the weights of your rolls of film. Also periodically check core weights. Weigh 10 cores and divide total by 10 to obtain a fair average. Subtract core weight from roll weight to obtain chargeable film weight. Some suppliers have been known to invoice weights greater than actual film weights or reduce film weights on film sold by the roll.

MAINTENANCE

Blown stretch film is coated with a “tackafier” to cause “cling” to overlapped film when applied to your pallet load. This tackafier can be deposited on your pre-stretch rollers and in time reduce the amount of pre-stretch produced by your control setting. A reduction in pre-stretch increases the amount of film used and in turn increases your cost. Check with your film and machine representative on any maintenance procedures required.

LOAD COVERAGE

To properly contain your product load on a pallet, the following should be basic procedure:

1. Include at least the upper 2 or 3 inches of your pallet in the first 1 to 3 wraps of stretch film. A ½ inch single wing pallet will improve the holding power of the film both vertically and horizontally to the pallet. If the sharp corners of the front and

rear deck boards tear the film, blunt the corners with a hammer.

2. Overlap the spiral 3 to 5 inches. Check your machine supplier for their recommendation. Note: the more the overlap, the more the cost of the film. If overlap is too little, the retention of the load is jeopardized.
3. Permit spiral to rise above the load at least 3 to 5 inches. The film above the load will shrink inward which will tie down the load vertically. If small products are in the load and pilferage is a problem, place a corrugated or equivalent sheet on top of the load. The sheet dimension should be close to the load length and width dimension to make unauthorized removal difficult and detectable.
4. When stretch wrapped products exceed the length or width of the pallet, film on adjacent loads in a truck or trailer can "cling" to each other and cause problems in removal. Check with film supplier and your trucking company. Remember: the trucking company wants no problems and the film company wants more sales dollars. If the problem exists, film with tackafier cling on the inside only is available.

SPIRAL ROLL WIDTHS/MACHINE

On pre-stretch machines, the roll width can be specified. This will determine the maximum roll width but not the minimum width. If your product is better serviced with two roll widths, specify the greater width for your pre-stretch. This will only work if the roll is placed over the spindle <u>and down</u>. This causes bottom of roll to be at same starting point on pallet regardless of roll width. Conversely, pushing the roll up onto the spindle raises the starting point of the smaller roll a distance equal to the difference on roll width. Example: A 20-inch roll on a 30-inch pre-stretch machine will have the bottom of the roll approximately 5 to 7 inches above the pallet. This concept is based on significant usage for multiple roll widths and only one justifiable machine. Note: a 30-inch roll cannot be used on a 20-inch pre-stretch machine.

MACHINE CHARACTERISTICS

Unless significant improvements have occurred, the following is recommended:

1. Dual chain drives on vertical lifting of pre-stretch component.
2. Chain drive to turntable.
3. Four or more support wheels under turntable.
4. Secure location of control settings is desirable.
5. Dust free location of control settings.
6. High or low profile is your choice.

There are a variety of machines on the market. The simplest is a unit with a motorized turntable, a brake on the film roll and a manual crank to raise and lower the film in a spiral mode. Pre-stretch, powered spiral lifting and lowering, automated features, and conveyorising are available at additional costs.

CHAPTER 9

HAZARDOUS MATERIAL PACKAGING CONSIDERATIONS

●●

IN THIS CHAPTER

●●

●●

HAZARDOUS MATERIAL PACKAGING CONSIDERATIONS

This section is not for the "faint of heart". Packaging hazardous materials is very serious business. Failure to properly package, test, document, store, transport, mark, and display warnings can cost you personally $25,000 in fines and a five-year jail sentence. If you are involved with manufacturing, purchasing, returning, or handling hazardous materials, I strongly suggest that you read this section slowly and repeatedly. Ignorance of the regulations pertaining to hazardous materials (HAZMATS) is not an acceptable excuse. If you fail to fully comply, you will be in deep trouble.

What is a hazardous material? The "Code of Federal Regulations (CFR) part 171.8, is defined as the following: "A hazardous material means a substance or a material which has been determined by the U.S. Secretary of Transportation to be capable of imposing an unreasonable risk to health, safety, and property when transported in commerce and which has been so designated. The term includes hazardous substances, hazardous wastes, marine pollutants, and elevated temperature materials…materials designated as hazardous under part 172.01…and materials that meet the defining criteria for hazard classes and divisions are contained in part 173."

Note: A current copy of CFR 49 should be maintained at all times by any person or company shipping or receiving HAZMAT materials.

There are companies that, for a fee, will provide updates to CFR 49 on a monthly basis. You also can access the D.O.T.'s own website for very extensive coverage of HAZMAT regulations, ongoing proceedings, updates, etc., but the current regulations in book form is a better shop floor and office reference. D.O.T.'s website currently is http://hazmat.dot.gov. Proposed regulatory changes and final rulemakings also are printed in the Federal Register under "Research and Special Projects Administration (RSPA). Persons qualifying as "HAZMAT employees", as defined in part 107 of CFR 49 must receive special HAZMAT training sanctioned by the Federal Department of Transportation (D.O.T.) and at a depth commensurate with their responsibilities.

The Federal government, again, prescribes the regulations in CFR 49 for transporting "HAZMATS" by all modes of carriage in interstate commerce. Interstate commerce includes commercial transportation between states via any mode, transport via inland waterways, and transport between the U.S. and other countries. States can also regulate intrastate transport of these materials up to a point but generally accept and defer to federal regulations. Enforcement may be federal or state and penalties for infringements either civil or criminal, depending on the nature of the offense. D.O.T. regulations incorporate the United Nations "recommendations" governing the labeling and packaging of dangerous goods for transportation in international commerce so these packages intended for HAZMATS are now called "UN" packages. Most countries also use the UN's "recommendations" as their foundation for their own HAZMAT regulations; however, they also make their own laws governing them.

Some transportation services such as U.P.S., steamship lines and air carriers have their own regulations, which can and do extend beyond the "basics" found in federal or state statutes. These must be observed where applicable. It stands to reason that no company shipping or arranging shipments for itself or others by ocean or air should be without current editions of their respective industry regulations covering HAZMAT packaging, marking and documentation. These are found in the International Maritime Dangerous Goods Code (IMDG) for ocean shipments and the International Civil Aviation Organization (ICAO) for air shipments. Note: The IMDG regulations are organized into several loose leaf binders for easy revision whereas UPS and the air regulations are reissued annually in bound book form.

HAZARDOUS MATERIAL IDENTIFICATION

It is very important that persons involved in handling, packaging, preparing documentation, and shipping products or materials follow a procedure similar to the following:

1. Determine if the product or material is hazardous per D.O.T. regulations.

2. If hazardous, what is its proper shipping name, the United Nations ("UN") number, hazard class, and packing group number?
3. The correct packaging required.
4. The correct labels and markings need to be applied to the package.
5. Are there any additional requirements due to type of transportation?

Information relative to (1) through (4) above is found in CFR 49 part 172.101's Hazardous Material Table. Information needed in (5) could be found in special air, water or carrier organizations previously mentioned. The 172.101 table also contains special provisions, packaging exceptions and quantity limitations by HAZMAT material and mode. This can be very confusing but the fact is that everything in the table, every note or reference mark, needs to be checked and reviewed for application to your product. You are responsible. It is important to re-emphasize that D.O.T. prescribed training covering using the CFR and all other aspects of hazardous material shipping must be provided to all employees responsible for shipments. It is required by law and it protects the company and its employees. Instructions regarding use of the CFR 172.101 table and classification of hazardous materials are taught in depth in the aforementioned classes.

Helpful tip. Not sure if a product or material is hazardous by law under 172.101 because you do not know its proper shipping name or

UN number? If your company manufactures it, get a determination from a member of the product engineering staff familiar with the item after he's gone through the 172.101 listing with you. Do not make a guess on your own. If you buy it from a supplier, request a Material Safety Data Sheet (MSDS) from the supplier or your quality assurance department. If the product is hazardous, it should be described as such on the MSDS and reference both the "proper shipping name" and its UN number. This will get you into 172.101. It may also provide you with information in other sections that can be helpful in its classification, packaging and marking. Always be thorough when using an MSDS. Remember that you, the shipper, are ultimately legally responsible for HAZMAT shipments leaving your facility. You must satisfy yourself that the information you have been given or provided is correct. WARNING: There is no guarantee that markings and/or data on incoming hazardous materials is correct, especially if it is from overseas. If the information is still suspect, make your management aware that you are not able to prepare the item for shipment and the reason why. Pay special attention to whether the product, as packed upon receipt, can be repacked in its packaging and reshipped by the same or other mode of transport. A very common misconception is that HAZMAT packaging is universal to all modes and may be resealed and reused.

HAZMAT-trained employees can navigate through 172.101 successfully and will know how to find and use the proper shipping name of a hazardous material and how to determine its legally acceptable packaging by "mode of transport." They will also be able to find out how to mark and label the package, its maximum legal quantity per package, and how to test it to meet packaging material specifications and UN performance tests. These tests and specs control the quality of components in the package, its design and the overall performance of the package under specific test conditions. These test procedures and conditions are contained in CFR 49 part 178 along with package certification/marking instructions governing construction, materials and testing for HAZMATS shipped in non-bulk packaging. We are concerned with non-bulk packaging as defined in CFR 49, which is further divided into three types:

1. Single packaging (example: sealed can tested, properly marked and able to ship as is)
2. Composite Packaging (example: rigid outer container with internal liner either attached to outer container or takes on its shape. Liner itself cannot be transported safely without protective outer container.)
3. Combination packaging (example: strong corrugated outer package housing multiple rigid plastic bottles containing compatible hazardous materials)

Legal definitions of these three non-bulk hazmat packagings are found in CFR 49, part 171.8.

The sections of the CFR referenced in part 172.101 will indicate which type is authorized for the HAZMAT being shipped. HAZMAT packaging selection considerations:

1. All references, notes and symbols in part 172.1011 for the commodity being shipped must be checked against your shipping and packaging application.
2. IMDG, ICAO, and UPS guides must be checked for other requirements if involved in shipment.
3. The commodity must be physically and chemically compatible with the authorized HAZMAT package. What does this mean? The CFR's authorized packages are often described in general terms to cover a wide variety of applications across the hazard classes. For example: an authorized packaging description may state that plastic liners are required in construction but <u>do</u> <u>not</u> identify the plastic material, its thickness or the closure method. It is up to the user to determine the answers through his knowledge of packaging materials, the design process and part 178's performance testing.
4. The actual transportation and material-handling environment must be considered if authorized packaging options are available. Higher levels of protection for very rough handling or outdoor exposure exist.

HAZMAT PACKAGING DESIGN & SOURCING

Let us assume that researching part 172.101 has led you to a package type applicable for the HAZMAT you are shipping. You have addressed (1) through (4) above and are satisfied that you can move into sourcing it. There are two options: (a) design, test and arrange for its production yourself or (b) buy an off-the-shelf UN tested package suitable for your product in the state (solid, liquid or gas), quantity and by the mode you wish to ship it. Consider doing it yourself first...Who is doing the design work? You, a staff engineer, an outside manufacturer of corrugated boxes or other packaging materials? What are your and their qualifications and experience in dealing with HAZMAT materials? Many suppliers of packaging boxes and materials want nothing to do with them because they automatically become a possible defendant if and when something goes wrong. Did the supplier test his product or material for use when packaging HAZMATS and follow your guidance and specs? You need to confirm that.

The design process involves more than the selection of appropriate materials to contain and protect the product, how they will be made and used, who will make them, how the package is assembled, how it will perform in UN tests and in the field use and how much it will cost. You also need to consider how storage and recycling concerns may impact the design and how components are managed. (This is true whether you design it internally or source finished packages from suppliers.)

It is suggested that companies without in-house expertise to do this work seek the services of packaging consultants with experience in successful UN performance packaging design and development. You also need to select a D.O.T. certified lab with the appropriate capabilities to do your testing for you. Pick one easy to visit who may already have experience with your product or industry. You should be wary of advice from your company's traditional packaging suppliers regarding their design capabilities and willingness to develop a design for HAZMAT products or materials for you. Of course, there are exceptions to this. Be careful and cautious.

Buying "off the shelf" makes sense if the authorized HAZMAT package is commercially available for your product now, is available on short order and its use level is expected to be low. Off-the-shelf UN packages are commonly available in many variations through safety product catalogs and industrial supply houses. The testing of these HAZMAT packages has already been done. They probably have a good history of use. One negative feature is that they are multi-purpose and may not match your volumes or weights exactly. The key to using them is that you must screen them against D.O.T. and other regulations applicable to your product to be absolutely certain of their application to your purpose. Contact distributor or manufacturer and ask questions. Make a list of them first and don't settle for general answers or claims. You also must be certain in advance that your company's employees have the ability to assemble and properly mark the package. Special sealing or torquing equipment may be required for caps and other closures. It is better to know this before buying the special package. Note: UN package assemblers must follow the manufacturer's instructions perfectly every time otherwise the package will not perform in accordance with its design and your company will face serious consequences if it should fail. This applies to internally designed and sourced UN packages as well as those purchased off-the-shelf.

CONTROLLING HAZMAT PACKAGING

Let us assume that you have introduced a UN tested HAZMAT package into your inventory. It is not just another standard corrugated or plastic container and cannot be treated as such because of the legal liabilities that affect employees and the company when it is put into service. It is also usually much more expensive. You must control how it is going to be stored and used by your company to assure that it will keep your product safe, legal and protected when transported. There are four basic control considerations:

1. Establish a document trail with suppliers. Packaging components individually must meet all applicable performance criteria mandated in CFR 49 for the life of

the package. Make sure you have MSDSs or other certifications from each manufacturer that confirm that they are being met.

2. Keep UN performance tests for the finished package current and available for D.O.T. inspectors. CFR 49 part 178 prescribes applicable performance tests by package type and its retesting schedule. Establish a re-test reminder file.
3. Formally control all modifications to UN tested packages. You are legally required to build the UN or HAZMAT package exactly the same as it was tested. The only valid modification to a UN-tested HAZMAT package is a reduction in the amount of the hazardous material it was tested to carry providing that any void space created is filled with inert material. Don't make assumptions that other HAZMATS can use it unless you planned for them to be used from its inception.
4. Issue assembly and storage instructions to quality, material control and manufacturing managers so they can take steps to address its safety and legal sensitivity with their respective supervisory staffs. Make sure your vendor's MSDSs cover storage conditions and anticipated useful shelf life. It may be advisable to negotiate storage and "just-in-time" arrangements with vendors who can store for you under ideal conditions rather than tempt fate by doing it yourself. Keep in mind that a package

manufactured in 2001 will carry that date in its UN marking code. Will you, your customers and D.O.T. inspectors have confidence in its performance if it is not used until 2006? Develop a disposal or shelf life procedure to address this and make certain that your quality staff actively inspects any stored HAZMAT packaging for changes in its state.

RECYCLING/DISPOSAL

Some HAZMAT packages also may be refurbished after use and restored to service. Most of these involve products shipped in bulk, intermediate bulk or drum containers_beyond the scope of this discussion. In non-bulk packaging, it is important to know that paper based packages cannot be reused or refurbished. You must know your package's limitations regarding its reuse capabilities. It is up to the conscientious shipper to provide information to its employees, customers and potential end users regarding disposal or recycling used HAZMAT packaging. This may take the form of advisories printed on the package or inserted into it addressing its disposal or suitability for reshipment by specific mode of transport. This also applies for the HAZMAT materials that you receive. Are you receiving useful information from suppliers now? Some of it actually may be legally required on or in HAZMAT packages carrying products sold or delivered directly to consumers under "right-to-know" consumer protection laws. Persons responsible for this kind of packaging must become familiar with the applications of warning labels and other advisories required under federal and state consumer protection regulations.

LAST THOUGHTS...

PACKAGING MARKINGS...To insure a UN package's compliance with test regulations, a unique code is imprinted on one of its visible surfaces. Its composition is governed by CFR 49, part 178.503. This code tells trained inspectors the packaging performance standard under which the package was successfully tested, its date and origin of manufacture, and the identity of the testing facility that tested and approved the package. If you do your own testing, your company name and address becomes the identified tester. Carriers can refuse to handle the package if code is not on shipping container. If you are buying hazardous materials, make sure that your receiving and quality control staffs are aware of and look for these marks on products being delivered to assure that they give them careful handling to avoid accidents. Likewise, if your company over packs packaged HAZMAT products, D.O.T. regulations also prescribe the markings and labels required on the exterior container and personnel in shipping capacities must mark them accordingly.

DON'T FORGET MAILROOM STAFF...Mailroom staff may become "HAZMAT employees" and need D.O.T. HAZMAT training.

Samples of HAZMAT materials may be received at administrative offices for various reasons including inspection and analysis by executive, purchasing or engineering staffs. When the package is opened, it should be considered compromised and cannot be reused even if it is a UN-tested package. There are exceptions to this, of course, and only a D.O.T. trained employee will be able to determine this. An un-opened UN tested package also may be suitable for ground transport only based upon its testing and not be legal for another mode, especially air transport. This is very common. It also is common for engineers, especially, to seal and reship these products to other sites for further inspection and trials. Trained mailroom staff confronted with these situations can take appropriate action to get the right people involved and eliminate potential liabilities that may arise.

Are you confused by what you have read on HAZMAT packaging? What you have read is the important highlights. Many, many additional pages would be required to cover the subject in depth. Get management to authorize training. Sign up for D.O.T. courses. You have already been told the penalties you are exposed to if you make a mistake. Again, HAZMAT packaging is a serious business. Treat it as such.

To give you further insight as to the complexities of HAZMAT Packaging, the following website is listed for up-to-date information http://hazmat.dot.gov./doc.htm.

Label and Marking Instructions

- Required printing includes:
  - **UN Certification Marking** — **Hazard Warning Label**
  - **Proper Shipping Name** — **Identification Number**
- Application of the certificate marking is the box manufacturer's responsibility.
- It is expressly prohibited to mark any package with a certification marking if the package does not meet the specified instructions.

UN Certification Marking Example:

 **4G/Y25/S/93/USA/+AA0000**

4G Packaging identification code: 4G for fiberboard boxes

Y Performance standard for which box was tested: X, Y, or Z
25 Mass in kilograms for which the package design has been tested
S For solid or inner containers
93 Last two digits of the year of manufacture
USA Where the container was manufactured and marked
+AA0000 Name and address or symbol of company responsible for compliance

Proper Shipping Name (PSN)
- Required for all hazardous material packages
- Must be near the hazard warning label

Identification Number
- Must accompany the PSN on all UN performance packages except those for limited quantities and consumer quantities

Hazard Warning Label
- Diamond shaped
- Correct color and picture according to DOT instructions
- Size about 4" X 4"

Additional Label Requirements

Appearance

- Letter size must be .5" except for packages less than 8 gallons or 66 pounds, then must be .2" high.
- Color of markings must contrast the package background and must be permanent or non-removable (such as labels with non-removable adhesive).
- Must be on one side of the box, except for boxes that will be palletized or unitized, then must be on two adjacent sides.

For Liquids

- Must have package orientation labels (red or black) that are on two opposite sides of the box. They must be at least 74mm X 105mm in size.

For Boat Shipments

- Must have *Marine Pollutant* label on packages that contain marine pollutants.

For Air Shipments

- *Cargo Aircraft Only* labels must be placed on packages that are forbidden for transportation on passenger carrying aircraft.

Overpacks

- Must be marked with a statement that the inner packings comply with the regulations. Example:
  INNER PACKAGES COMPLY WITH PRESCRIBED SPECIFICATIONS
- Must have Proper Shipping Name and Hazard Warning Label on outside.

CHAPTER 10

GOVERNMENT CONTRACT PACKAGING

●●

IN THIS CHAPTER

●●

●●

GOVERNMENT CONTRACT PACKAGING

Selling to the U.S. government can be very complicated. Many companies think that they can produce and package their product the same way they produce and package for their domestic customers. They have paid a heavy price for this error. When the government determines that the packaging provided does not meet their specifications, they can and usually will send your product to a packer of ***THEIR CHOICE*** and repackage it to the proper specifications and send you the bill. You will have ***NO OPTION*** but to pay the charges in full.

To protect yourself ***BEFORE YOU SUBMIT A BID***, do the following: thoroughly read the contract, especially the part that refers to packaging and marking; obtain copies of all specifications referenced as they are free. (There are locations scattered around the country that stock these specifications. You can fax your request or phone it in.) A specified specification can contain another required specification unbeknownst to you.

CAUTION: Many companies are requesting specifications for various contracts and can deplete the quantity in stock to zero. Not having the specification does not get you off the hook. You still must comply with its unknown contents. You may have a copy of a specification that is outdated. You do not know. The government will send you the latest revision upon ordering. Keep trying to obtain specifications that are "out of stock" before you bid the contract. Some contracts are not clear as to specific packaging requirements. If you are not sure, call the contracting officer whose name and phone number is on the face of the contract. Government packaging can fall into three categories:

LEVEL A

This is the most demanding level and requires maximum protection and preservation for the worst of conditions.

LEVEL B

This is a little less demanding than level A but still requires significant protection and preservation.

LEVEL C

This is basically commercial packaging, but there may be special demands.

We have just discussed packaging. We will now discuss packing. Yes—there is a difference! There are three levels of packing:

LEVEL A LEVEL B LEVEL C

The demand for protection and preservation closely follow the descriptions under packaging. Most contracts will spell out packaging/packing levels as: A/C, A/A, C/C, B/C, etc.

Packing is the unitizing of packaged products into a unit load which may be palletized. Packaged product can be single product packaged or intermediate package containing multiple packaged units.

INSPECTION/INSPECTORS

All government contract shipments will be inspected. Some will be inspected within your facility. Some will be inspected at first destination. The inspectors will follow the demands of the contract closely. Some inspectors, when reasonable, will grant you some latitude in interpretation of the specifications. Others will hammer you until they are satisfied.

All packages, intermediate packages, and unit loads must comply with marking requirements. This can be significant. Mil STD. 129 has been the bible on marking.

Warehouses or depots have very high ceilings with the result that your products can be stacked very high or be on the bottom of a high stack load. The ability of your packaged/packed product to support this load without compression damage is critical.

If you believe that the contract calls for excessive protection, you can appeal to the contracting officer. He will probably turn you over to a packaging specialist. Not all specialists are familiar with all commodities. Mistakes have and will be made in contract specifications for packaging/packing. You will have the problem of convincing the specialist. It may require a visit to his location. It has happened with a successful conclusion.

Contract quantities can be an annual quantity with many unknown small quantity releases throughout the term of the contract. How many packages do you buy or stock?

GLOSSARY

BOBST – Name of equipment (machine) manufacturer.

CIR-CUT TOOL – This is a steel rule cutting die mounted on printer/slotter machine to cut a shape into the corrugated board. Most common use is a "Hand Hole".

ECT – (Edge Crush Test) procedure. This is a test suggested by paper manufacturers to evaluate test value of corrugated board.

FOL – Full Overlap Flap Carton

FPC – Five Panel Folder

G – Force of gravity 32 feet/sec/sec

HSC – Half Slotted Carton (No bottom flaps)

RSC – Regular Slotted Carton (Length flaps meet in middle)

ABOUT THE AUTHOR

The author was first assigned a packaging project in 1958 with Industrial Engineering being his basic function. He went on to become a Packaging Engineer for a large electronic instrument company. He later designed and implemented packaging involving various materials for a division of a major communications company.

He entered the sales field specializing in various packaging materials. These materials are covered in this book. In 1982 he joined a major battery company. He rose to Corporate Manager of Packaging Design and Packaging Purchasing. In 1989 he started his own company specializing in consulting and sales. To this day, he has stayed involved in packaging. His experience covers 45 years.

www.ingramcontent.com/pod-product-compliance
Ingram Content Group UK Ltd.
Pitfield, Milton Keynes, MK11 3LW, UK
UKHW041937190726
13854UKWH00004B/1648